# DIGGING DEEPER

# DIGGING DEEPER

## MAKING NUMBER TALKS MATTER EVEN MORE, GRADES 3–10

RUTH PARKER &
CATHY HUMPHREYS

STENHOUSE PUBLISHERS
PORTSMOUTH, NEW HAMPSHIRE

Stenhouse Publishers
www.stenhouse.com



Library of Congress Cataloging-in-Publication Data
Names: Parker, Ruth E., author. | Humphreys, Cathy, author.
Title: Digging deeper : making number talks matter even more, grades 3–10 / Ruth Parker and Cathy Humphreys.
Description: Portsmouth, New Hampshire: Stenhouse Publishers, [2018] | Includes bibliographical references.
Identifiers: LCCN 2018017098 (print) | LCCN 2018020296 (ebook) | ISBN 9781625312051 (ebook) | ISBN 9781625312044 (pbk. : alk. paper)
Subjects: LCSH: Mathematics--Study and teaching (Elementary) | Mathematics--Study and teaching (Middle school)
Classification: LCC QA135.6 (ebook) | LCC QA135.6 .P376 2018 (print) | DDC 510.71/2--dc23
LC record available at https://lccn.loc.gov/2018017098

Cover design, interior design, and typesetting by Tom Morgan Blue Design (www.bluedes.com)
Manufactured in the United States of America

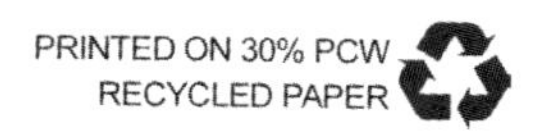


24 23 22 21 20 19 18 9 8 7 6 5 4 3 2 1

# Contents

# Acknowledgments

First, a message for Toby Gordon, our intrepid editor. Toby, how can we thank you? You believed in this book and made it better in every way. You were excited about what we were thinking, and you were right there with us, trying to figure things out. You knew when to push, when to offer encouragement, and when to give us space. Thank you for making this book happen.

We also want to thank the teachers and students whose voices you hear and faces you see in these pages. Jay Jahnsen, Nisha Puri, and Hailey Gilmore from North Thurston Public Schools in Lacey, Washington, thank you for sharing your teaching practice in such a public way. And thank you, Emily Aufort, for arranging and coordinating our filming efforts so seamlessly. A big thank-you also to Melissa Johnson and Tara Palomares from Fremont High School in Sunnyvale, California, whose thoughtful reflections will resonate with so many teachers. And Cathy Young, we love knowing that your work will live on.

A special thanks also to Debbie Olson, Patty Lofgren, and Lisa Mesplé, our colleagues at Mathematics Education Collaborative. Our many conversations with you have expanded our thinking and supported us every step of the way. We also want to thank the teacher leaders who participate in the Washington State Number Talks Leadership Academy. Thank you for welcoming our partially formed ideas, trying them out in your classrooms, and reflecting with us along the way. Jamie Souhrada, we especially want to thank you for testing our new wait time ideas with your students and sharing the results with us all.

The voices of countless friends, colleagues, and students were in our heads as we were writing this book. We can't begin to name you all, so we won't even try, but we're aware of how fortunate we've been to have so many treasured and wise teachers. Know that we are filled with gratitude for each of you.

# Introduction

When we wrote *Making Number Talks Matter* (Humphreys and Parker 2015), we knew that there was still more to learn about how to help teachers bring Number Talks to life in their classrooms. Almost as soon as the book was published, we started finding issues we wanted to address with greater clarity and depth; we also began to think about and discuss new ideas that were just coming onto our radar. This continual process of questioning and revising our own practice is central to who we are as educators, so it isn't much of a surprise that now, three years later, *Making Number Talks Matter* has a companion.

In *Digging Deeper: Making Number Talks Matter Even More*, we share the insights and revised thinking that have emerged from our ongoing experiences with Number Talks—our own and those we've watched—over the past few years. This book is not meant to stand alone as a reference for how to enact or plan Number Talks. Rather, we wrote it to enhance the ideas in *Making Number Talks Matter* by digging more deeply into ideas and events that have come to our attention as being tricky, problematic, or just plain hard. Each chapter addresses teaching practices we want to get better at, ideas that we have developed new perspectives on, or issues we have become concerned about. Where possible, we have included a vignette, a video, or both to illustrate a dilemma, a question, or an idea that caught our attention. These videos are intended not as exemplars but rather as windows into the ideas we address throughout the book. The teachers you'll see on the videos have given us all a gift by opening up their practice so that we can learn from it. We asked Hailey, Nisha, and Jay, the three classroom teachers whose Number Talks are part of this book, to briefly introduce themselves to you.

## HAILEY GILMORE

**I am a National Board Certified Teacher and district-based math instructional specialist. Since being introduced to Number Talks in 2014, I've led Number**

**Talks in many kindergarten to fifth-grade classrooms and math intervention support groups. I enjoy collaborating with teachers as they develop Number Talk routines in their classrooms, and the students continue to amaze me as I listen to them explain their thinking and watch them grow as mathematicians. Currently, I am participating in the Number Talk Leadership Academy project in Washington State.**

### NISHA PURI

**I'm an eighth-grade teacher and have been teaching for nine years in Washington State. About five years ago, I was blessed to join a group of teachers who would spend their summers working to strengthen and deepen their own fragile understanding of mathematics. The first summer, I learned about Number Talks, and I started using them in my classroom that fall. I marveled at the various ways students tinkered with numbers. Number Talks translated into students growing more confident in their abilities and into a willingness to try new strategies.**

### JAY JAHNSEN

**I'm a National Board Certified Teacher working at River Ridge High School in Lacey, Washington. I've been teaching for ten years and was first introduced to Number Talks a few years ago after hearing my coworker Jami talk about this new, cool presentation she saw at a professional development workshop. I'm a firm believer that activities in math classes should provide opportunities for all students to make sense of their learning. Activities should be challenging to students who are already successful in math, and still both challenging and accessible to students who struggle. Number Talks meet all these requirements. My class that you will see is a very "normal" high school class with a wide variety of ages, backgrounds, abilities, and income levels. I hope you will find Number Talks to be as valuable in your classroom.**

We also want to introduce Melissa Johnson and Tara Palomares, who, as high school preservice teachers in the Stanford Teacher Education Program, participated in a five-week collaboration to plan and enact ten Number Talks. Their work provoked and enriched our thinking about Number Talks in high school for *Making Number Talks Matter*. And although we aren't able to share the Number Talks themselves, we include video excerpts of interviews with Tara and Melissa at the end of the project. After leaving Stanford, Melissa taught geometry and AP calculus at a public high school in Littlerock, California, and currently teaches in a virtual high school. Tara teaches geometry and pre-calculus in Albuquerque, New Mexico.

What follows is a brief overview of each chapter.

Chapter 1, "A High School Number Talk: The Case of Frosty," sets the stage for the subsequent chapters in this book. Shown in its entirety, this Number Talk is as unpredictable and full of dilemmas as any that a teacher might face when working with students' ideas, no matter the grade level. The problem with watching videos of Number Talks, though, is that while we can generally see what teachers do, we aren't privy to how they *decide* what to do. Their decisions, based on what they *notice* combined with a swirling mix of their own beliefs, attitudes, and habits, are hidden from view. So Cathy sets out, in this chapter, to make her teaching decisions transparent. Through the lens of her goals for the lesson, she describes how she makes sense of what she notices and how she decides what to do. This chapter highlights how complex, but also how stimulating and downright fascinating, the deceptively straightforward fifteen minutes of a Number Talk can be.

Wrong answers and mistakes—how we think and feel about them and what to do about them—is at the heart of teaching. In *Making Number Talks Matter* (Humphreys and Parker 2015), we discuss how our role as teachers during Number Talks is not to "fix" students' thinking in the moment. But are we still teaching if we don't point out mistakes and show students how to correct them? In Chapter 2, "Mistakes," we explore different ways that teachers deal with mistakes and confusion during Number Talks and consider what it can mean to use mistakes as sites for learning.

Chapter 3, "Revising the Number Talk Routine: Rethinking Wait Time," explores some new-to-us ideas about wait time during Number Talks. In *Making Number Talks Matter*, we discuss "being comfortable with plenty of wait time" (2015, 18), which is widely acknowledged to support student learning. As we visited classrooms, viewed online videos of Number Talks, and watched each other's Number Talks, however, we noticed that it's not that simple. While teachers generally allow plenty of wait time during Number Talks when students are figuring out their answers, most of us tend to move fairly quickly from one student's strategy to the next with little time for students to absorb and ponder those strategies in any meaningful way. We began to wonder where else we could be more purposeful about waiting during Number Talks. Where else should we be more purposeful about waiting? And why? Our new thinking around the potential for wait time led to the revised Number Talk routine in this chapter.

How to ask questions during Number Talks is a concern that arises frequently in our discussions with teachers. In *Making Number Talks Matter*, we wrote, "Through our questions, we seek to understand students' thinking" (2015, 26). But this principle is perhaps too general to help teachers know what kinds of questions to ask. In Chapter 4, "The Questions We Ask: What? Why? When?" we offer specific guidance on the kinds of questions teachers can use during Number Talks to help elicit students' thinking. We suggest questions to avoid and discuss how our questioning can invite student engagement.

Another difficulty many teachers face is students' lack of experience with generating their own mathematical ideas. Teachers ask us, "How do we get our students to come up with new strategies unless we show them?" Certainly, the persistent (and misguided) notion that there is one best way to solve any arithmetic problem permeates students' thinking. Even when students "know their facts," they may never have had the chance to "play" with numbers and operations. Some students may be "stuck on the algorithm" and have trouble breaking away even when it doesn't work very well for them. Other students don't even understand why they should try something new when having one way that works is what they have learned to care most about. Chapter 5, "Nudging: Helping Students Have Their Own Mathematical Ideas," presents our emerging ideas about how to encourage students to access their own understandings and ideas when they are given a new problem to solve.

Chapter 6, "Dot Talks," addresses how to optimize the use of Dot Talks to engage all students and to shift the classroom culture to one of a learning community. We focus on practical suggestions for getting Dot Talks started and reasons for keeping them going. The chapter also includes specific guidance in response to questions teachers have asked us about Dot Talks. Teacher moves during Dot Talks and Number Talks are almost the same, so we focus throughout the chapter on the role of Dot Talks in setting the stage for Number Talks.

What does it mean to have a "safe" learning environment? In *Making Number Talks Matter* we talk about the importance of creating a learning environment "where all students feel safe sharing their mathematical ideas" (2015, 28). One teacher, though, said to us, "Of *course* everyone wants students to have a safe environment for learning." We've come to realize that "safety" isn't easily defined and there are different beliefs about what makes a learning environment safe. In Chapter 7, "Safety," Ruth revisits the personal experiences that led to her beliefs about safety and shares her thinking behind the decision to always put students in control of whether and when they talk during Number Talks.

The need for Chapter 8, "More Bumps in the Road," was sparked by questions and concerns we hear from teachers who have been using Number Talks for a while. In *Making Number Talks Matter*, we shared ideas for getting beyond some typical bumps in the road that teachers experience as they get Number Talks started (2015, 163–75). In this chapter, we address teachers' questions and concerns about making Number Talks more meaningful.

Finally, in the Afterword, we summarize the many ways that Number Talks help students and teachers learn—together—to unsettle the prevailing culture of mathematics classrooms and to place sense making at the heart of mathematical activity.

We should also note that throughout the book, we talk about "many teachers," and we want you to know that we include ourselves in this group. We have experienced all of the stumbling blocks, difficulties, and challenges in this book, just as we're sure many of you have. Our suggestions throughout are based on our own personal efforts to find ways to get beyond the obstacles in order to expand the potential of Number Talks, making them a more vibrant, engaging, and meaningful part of mathematics classrooms.

CHAPTER

# 1 A High School Number Talk: The Case of Frosty

## Cathy's Reflections

### Setting

In October 2016, Ruth and I were invited to film Number Talks in an Algebra 2 class in a large public high school near Seattle, Washington. The teacher, Jay Jahnsen, had participated in a Number Talk institute during the previous summer and, enthusiastic about the potential of Number Talks, had been doing them since the beginning of the year.

The week before filming we visited Jay's class, which was composed mostly of sophomores and juniors with a couple of seniors. Jay introduced us to the students, talked about the upcoming video project, and then did two subtraction Number Talks for us to watch: 9.4 – 2.85 and 7.5 – 3.94. There was a friendly, relaxed feeling in the room, with students readily sharing their approaches. We noticed that most explanations involved removing the decimal points, calculating a whole number answer, and then putting the decimal points back in. One student used the open number line as a visual tool to add up from the subtrahend. Only one answer was offered for each of the problems.

We were particularly struck, though, by the supportive atmosphere Jay had nurtured so early in the year. Students made comments like, "I'm kind of afraid to do this, but . . ." and "I know what I'm doing, but I don't know how to say things." One student even said, "I stopped there, 'cause I got the wrong answer." It's rare to see teenagers talking so openly about not being sure or being wrong, and we've found that being willing to take those public risks is one of the intangibles that helps Number Talks transform math classes.

## Choosing the Problem

As I thought about what problem to do for my own Number Talk in Jay's class, I had several things on my mind. After subtraction Number Talks with decimals, I couldn't help but wonder what the students would do with fractions. And although I didn't know anything yet about what Jay's students understood about fractions, I did know from personal experience that students often come to high school with spotty understandings of what fractions mean, and avoid them whenever they can.

So I started thinking about a fraction subtraction problem that would be challenging without being too hard. At first, I thought about having them subtract a mixed number from a whole number (e.g., $8 - 5\frac{2}{3}$), but would that be too easy? And what different strategies might students use? The possibilities seemed limited, so, instead, I decided to choose subtraction of two mixed numbers. But what numbers, in particular, and why? This is a question I ask myself before every Number Talk no matter how long I have worked with a class. The particular numbers to choose for any Number Talk depend on what came before, what I've learned about the students, and what I'm wondering about.

Here is the reasoning process I went through for this Number Talk, with students I didn't know very well:

- First, I wanted the denominators to represent familiar fractions that students could more easily visualize, like halves, fourths, and eighths.
- Second, I wanted students to stick with fractions rather than convert them to decimals (which many students tend to do), so I wanted fractions whose decimal equivalents are less well known than halves and fourths. I thought eighths would serve that purpose.
- I wanted the fractions to have different denominators to give me better insights into students' understanding of fractions and to keep the cognitive demand high. But I also thought mental calculation would be easier if one of the denominators was a factor of the other.
- Finally, I wanted a problem that made the standard algorithm troublesome enough that students would need to reason their way through the problem. The easiest way I knew to do this was for the fractional part of the subtrahend to be greater than the fractional part of the minuend.

The particular numbers, then, didn't matter so much. I landed on 7½ – 3⅞, reasoning that this problem would be accessible in a variety of ways to students who understood equivalencies between halves and eighths (and maybe fourths).

Then I spent some time anticipating how I thought—or, maybe, hoped—the students might think about the problem. One worry I had was that they might get bogged down trying to use the traditional algorithm for subtracting fractions. Here is what I jotted down (see figure 1.1):

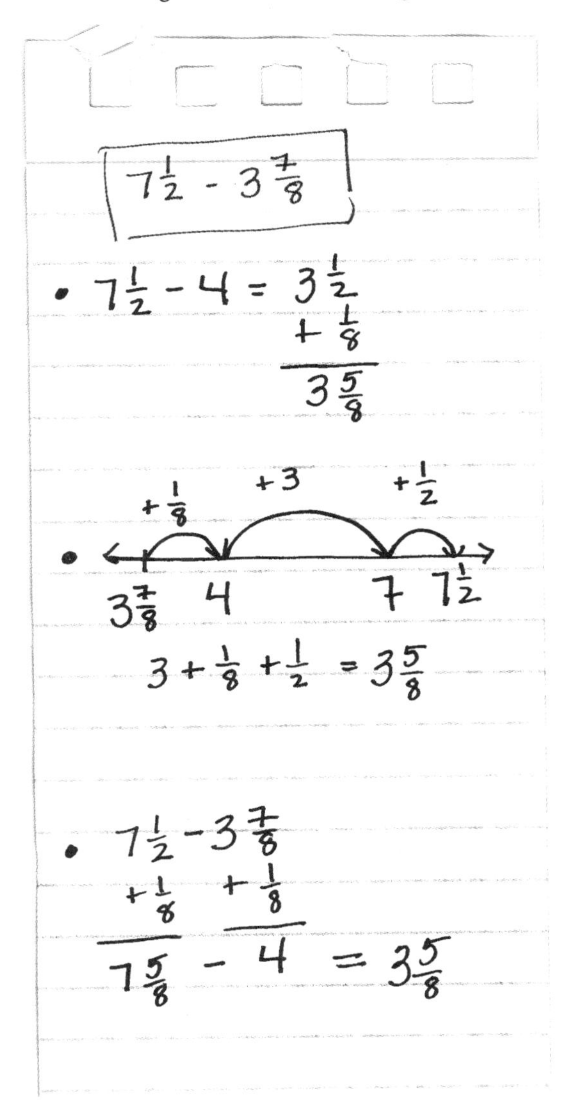


**FIGURE 1.1**
Photo of my notes.

Also in the back of my mind, as I planned the Number Talk, was an ongoing conversation Ruth and I had been having. We had become aware, across classrooms and also in our own work, that we craved more student interaction during Number Talks. Even when students were sharing a variety of strategies, we often felt a disconcerting "flatness" in the room. Something seemed to be missing, and that something, we thought, was more lively discourse around students' ideas. Maybe if students don't realize how important their ideas can be to other students' thinking, we need to be more explicit, as teachers, to help them with this. Therefore, I went in to do the Number Talk with two priorities in mind:

- Find out how students make sense of subtracting fractions.
- Try to get more students to engage with each other's ideas.

Here, we share the video of the Number Talk in its entirety—uncut and unedited.

Note: The audio was not connected—and the cap was on the camera lens—when I started the Number Talk, so for the first fifteen seconds you'll hear only my voice, starting midsentence, against the backdrop of a black screen. Also please notice that near the end of the clip, the video was turned off but the sound remained on. The video then resumes in time to capture an unexpected occurrence.

Subtraction of mixed numbers: a high school Number Talk.
http://sten.pub/dd01

## Reflecting on the Number Talk as It Unfolded

As soon as I mentioned fractions, a couple of students made a face (which I acknowledged). Not their favorite math topic! Nevertheless, I wanted to encourage them to "play around" and try to think about what made sense rather than remembering rules they had been taught. I put the problem $7\frac{1}{2} - 3\frac{7}{8}$ up on the board and waited and waited . . . and waited some more. Even after a full minute, there were very few thumbs up. What should I do?

Any classroom discussion has innumerable places where teachers need to make on-the-spot decisions that affect the direction of the lesson. We never know if the decisions we make are the "right" ones, or even if there *is* a right decision. Rather, we make our best guess about directions to go, depending on our priorities. I hit my first major decision point here.

*Choices as I saw them: Wait some more? Change the problem?*

Judging that more wait time wouldn't help many more students access the problem, I made a quick decision to modify it. Anticipating what problem to choose the night before paid off here because I could fall back on my first, discarded-for-being-too-easy kind of problem: a whole number minus a mixed number. So I decided to erase the $\frac{1}{2}$. I did warn them that I was going to change the problem, because sometimes students want to keep thinking about a challenging problem and just need more time. This day, though, there wasn't even a murmur of protest, so I erased the $\frac{1}{2}$. As I did so, I tried again to coax them towards "figuring out" as opposed to "remembering how." This time enough thumbs went up to make me think we would have enough ideas to work with, so I decided to go ahead and ask for answers, and I was glad that more than one answer was offered. As I told the students, it meant that we would have some things to talk about.

### Troy's Method

Troy started us off by justifying the (correct) answer of $3\frac{1}{8}$. Almost as soon as he started talking, though, his words and tone caught my attention. He said, "First, you would subtract 3 from 7, thereby getting 4." Using the word *you* instead of *I* made his strategy sound like a rule to be followed rather than simply the story of what he had done. So I tried to persuade him to speak as if it were his own method by asking, "Troy, when you said, 'You should . . . ,' what did you mean by that?" This question, though, only led him to repeat himself, so I resorted to rephrasing his reply, trying to emphasize that the method was how *he* had thought about it. I think this message probably escaped most of the students, but I was trying to communicate one of the essential ideas of Number Talks: the methods we use to solve problems come from our own personal ways of thinking about the numbers and operations involved. Those methods or strategies may indeed be a standard algorithm we have learned, but they become "ours" only when we make sense of why they work. It is that sense making that we try to support and communicate during Number Talks.

I could tell from the rest of Troy's explanation that he understood what he was doing—and why. But I wasn't sure who else understood why his method made sense, especially since there were other answers on the board (really only two different answers, because $\frac{39}{8} = 4\frac{7}{8}$). I also knew, from my experience as a middle and high school teacher, how jumbled the rules for operations with fractions can get in students' minds. And the faces of these students were so . . . blank. What to do? Here was another decision point.

### *Choices as I saw them: Thank Troy and move on? Have the class explore his method further?*

Troy's clear (to me) explanation and correct answer set off flashing caution lights in my head. Because it was the very first explanation, I knew that how I dealt with it would set the tone for the rest of the Number Talk. Indicating in any way that his answer was correct could silence the voices of those who had a different answer and perhaps keep the misunderstandings about fractions hidden from view. Besides, I was curious about how other students made sense of what he had done, and since one of my priorities was to engage students around each other's thinking, I decided, on the spot, to dig. I've found it sometimes helpful, in my teaching, to have students try to put what others are saying into their own words. This can help clarify what they understand and where they have questions about a strategy. Besides, it can be useful for students to hear the same idea expressed in different ways; even small variations in words can make a difference. And while it's not essential that students understand every strategy, thinking about what others have done can help students "see" in different ways and shift the tone of classroom discourse from *presenting* to *thinking together*.

Making the decision to dig into Troy's strategy, though, was a trade-off; it meant I wouldn't have time for as many other students to share their thinking. The bounded nature of Number Talks means that tough choices about time are made every day; choosing one path means not choosing another. But here I was, with only this one chance to do a Number Talk. I had already seen a variety of students share strategies during Jay's Number Talks, so I decided that helping them learn to interact with each other's ideas would be my priority.

I also needed to consider how Troy might feel about having his thinking publicly analyzed. Ruth has talked about the importance of protecting students who shrink under scrutiny of their ideas. I didn't know Troy, but his explanation was expressed with such apparent confidence that it didn't cross my mind to check to see if it was okay with him first. This might not have been fair to Troy.

Students restate Troy's strategy.
http://sten.pub/dd02

After I asked, "Can anybody put into words what Troy was just saying . . . what his strategy was?" first Josie and then Xander volunteered. Xander's choice of words, though, caught my attention: "I *think* what *they're* doing is . . ." In saying, "I think . . . ," Xander communicates the tentativeness and in-process thinking we hope to nurture in class discussions. Saying "I think . . ." leaves the door open for others to think, too.

Josie chortled, then, when I asked Xander to check directly with Troy to make sure he had accurately described how Troy had thought about the problem. I added a comment that, given the chance to do over again, I probably would express differently. Saying, "I really, really like it when people talk with each other and you check with each other" was my less-than-perfect way of communicating that one of our goals in math is for students to feel free to interact directly with one another. The sentiment is good, but I wish I had explained *why*: when students talk directly to each other or check with each other or ask each other questions—rather than waiting for the teacher to call on them—they are more engaged and everybody learns more.

At this point, I attempted to clearly record Troy's method, purposely avoiding the way subtraction of mixed numbers is typically shown in books. Josie, Xander, and Ashley had contributed their ideas, and Troy himself had stopped using *you* and started using *I*. Finally ready to move to another method, I called on Paul.

## Paul's Method

Paul uses the open number line to subtract.
http://sten.pub/dd03

Paul, in adding from 3⅞ to 7 on the open number line, used one of the strategies I had hoped students might think to use (see Humphreys and Parker 2015, 39). An elegantly simple way to visualize subtraction as distance, it's an example of how a strategy for subtracting whole numbers can also make fractions much easier to think about. After he explained his method, though, I wasn't quite sure what to do.

*Choices as I saw them: Thank Paul and move on? Ask a question that focused on some aspect of the strategy? Get students to talk to one another?*

Ah, hindsight! While I had anticipated the strategy, I hadn't thought enough about what I could do if it came up. Calling attention to this model for thinking about fractions could help more students shed the constraints of memorized procedures that, like cobwebs, clutter up mathematical sense making. Paul understood why his method worked, but did others? In retrospect, I wish I had just waited. Would someone have asked a question? Made an observation? One of the things I'm learning, as we discuss in Chapter 3, "Revising the Number Talk Routine: Rethinking Wait Time," is how powerful those quiet spaces can be. Instead, I filled the space with, "Why did you start at 3⅞ [*sic*]?" This didn't give us much, mathematically.

There are tremendous internal pressures on teachers to fill the silence and keep things moving during class discussions. And while a well-crafted question can highlight important mathematical ideas, if we truly want our students to be active mathematical collaborators, we need to let ourselves off the hook for being the sole stewards of their thinking. Students can share the responsibility for everyone's understanding if they believe this to be part of their position and mission as students—and if we believe it, too.

My next attempt to get students to engage around others' ideas also fell flat. I had already asked—twice—for students to explain another student's strategy, and it was getting old. This might have been a place to have students talk together about Paul's strategy. Or, perhaps even better, to let the idea sit and simply move on to someone else's thinking.

Finally, I asked if anybody had a different strategy for this problem, which led us to Frosty (his real name).

## Frosty

Frosty.
http://sten.pub/dd04

I was blown away by what Frosty said: "Can I ask a question, though? Why is all this other stuff, like, 3s? 'Cause 7 minus 3 isn't 3 — it's 4. And why is it ⅛?" A few hands were already up by the time he finished his question, but I asked everyone to put their hands down. Resolving the question itself was a lot less important to me, right then, than acknowledging Frosty's contribution—his gift, really—to the class.

Frosty's pluck, in the face of two strategies that exhaustively justified an answer that was different from his, was remarkable. He wasn't asking why his answer was wrong; he simply couldn't understand, despite the explanations, *how the other answer could be right*. Here was a student advocating for—even demanding—his right to understand. It's one thing to say, "It's okay not to understand!" but entirely another to have a classroom community that supports questioning and wondering as readily as it honors knowing. Jay and his students had already made it safe for Frosty to take this very public risk. Now I wanted to be explicit that what Frosty had just done was important for everybody's learning.

It's important for people to speak up when we don't understand, like Frosty has just done.
http://sten.pub/dd05

As I was speaking, I kept my eye on how Frosty was handling the attention. With a different student, I probably would have asked if he was okay with having others consider his idea, and maybe I should have done that with Frosty, too. But he seemed not only at ease but genuinely puzzled—even frustrated—and he didn't seem to be enjoying the attention as an end in itself. I'd been watching the clock, too, and knew there were only five to ten more minutes before the bell rang. In the back of my mind, I knew Irene was still waiting to share, but I also knew that her answer ($^{39}/_{8}$) was equivalent to Frosty's. I decide to plunge ahead with Frosty and asked him to say his question again.

Frosty restates his question.
http://sten.pub/dd06

But now what? Questions like this are catnip for teachers. It's hard to resist jumping in to ease our students' discomfort and (trying to) resolve their confusion; after all, this is what "teaching" has meant to so many of us for so long. But Number Talks are about uncovering students' ideas, in all their complexity, rather than smoothing over the rough spots. Besides, Troy and Paul had just given clear (to me) explanations that had left Frosty unconvinced. Why would my own explanation be any different? Jumping in myself simply wasn't an option.

"Why are you guys getting so many different answers?"
http://sten.pub/dd07

"Why," he said, turning to the class, "are you guys getting so many different answers [from mine] when it's, like, two easy numbers and there's not another fraction?" Almost immediately, four students raised their hands to respond. But only four.

*Choices as I saw them: Call on one of the four students? Or get the whole class involved?* I could have—should have—taken a deep breath and waited for other students, but *wait time* wasn't yet on my radar as a powerful way to invite participation (see Chapter 3, "Why Wait?"). Nevertheless, whenever only a few students are willing to weigh in, my intuition says that something is wrong. And since engaging students with others' ideas was one of my main goals for this Number Talk, I decided to get the whole class involved.

I asked students to talk about one particular part of what Frosty had originally asked: "Where's the 3 [in the answer] coming from [if] 7 minus 3 is 4?" I hoped this was a good enough question to spark the students' interest. In retrospect, I wish I had first thought to ask Frosty to explain what he meant by "there's not another fraction." But, as hard as we try, it's impossible to always find just the right question or teaching move for an idea under consideration. Things move quickly, with lots of ideas to consider and multiple ways to proceed.

All of a sudden, as the students started talking, the class came to life. I drifted back and forth, catching bits of conversations, before calling the class back together. I had only given them about a minute (which seemed much longer) but was invigorated and gratified by the students' energy as they talked to one another. After thanking them, I shared something I had overheard Frosty asking while I was standing nearby: "Why is mine wrong?"

Being wrong is part of learning math.
http://sten.pub/dd08

Here, I wanted to explicitly *position* Frosty as competent (for a discussion of the relationship between competence and motivation, see Horn [2017]). Competence in math is traditionally associated with getting quick, right answers. Remarkably, during the previous Number Talk in Jay's class earlier in the day, Frosty had revealed that very belief. In the following video excerpt, we watch as Frosty explains how he used the number line to subtract 26 from 74: after adding up from 26 using four "hops" of 10 to get to 66, he tries to figure out how far it is from 66 to 74.

"Bad at math."
http://sten.pub/dd09

Frosty was laughing, as were others. But this is no laughing matter. The narrow definition of what mathematics *is* determines the limited view of competence internalized by so many of our students. If, instead, mathematics is truly believed to be an open, creative discipline that merits curiosity, exploration, and sense making, then there are many ways to be competent. I wanted to explicitly frame Frosty's actions—publicly exposing and wondering about what he didn't (yet) understand—as a genuine sign of competence in math. I wanted students to hear that expressing confusion, as Frosty did, is a strength that contributes to everyone's learning.

By now, Frosty was beginning to think that his answer might actually be wrong because of something Ashley had said about "seven and zero eighths" that made him see how "you could do it like that." But he still wasn't sure. Here was another decision point for me. There were only a few minutes left, and I still hadn't given Irene a chance to support her answer of $^{39}/_{8}$. But the class had just finished talking in small groups about Frosty's question, and we hadn't shared out. What to do?

*Choices as I saw them: Thank Frosty and the class and move on to Irene's method? Have a brief share-out of how people were thinking about Frosty's question before moving on to Irene's method?*

I chose, yet again, to turn the question back to the class. But just as I began to talk, Frosty interrupted me, asking if *he* could choose somebody. I try to be careful to make sure students are in charge of deciding whether or not they speak in class, so I said, "Well, I don't know—if they want to be chosen!" Frosty turned to Tyler, who, we found out later, had never before spoken in class. Tyler said, "I'll try, I guess." But I wanted to double-check: "Is that okay with you?" And when he said, "Yeah," I thanked him.

Frosty helps decide how to process the discussion.
http://sten.pub/dd10

Unfortunately, because Tyler started his explanation by saying, "7$^{3}/_{8}$" (instead of 3$^{7}/_{8}$), it was hard to follow his reasoning. In fact, both Tyler and the students who followed him struggled to explain where the 3 came from. Despite the collective confusion, though, I was nevertheless encouraged by how students built on one another's ideas in a kind of collaborative effort to explain. Ashley, for example, said, "I was just gonna go off of what Tyler said." And Josie said, "Like she [Ashley] was saying . . ."

The feeling in class, though, seemed to be shifting. As Ashley and Josie tried to make convincing arguments, it felt like there was pressure on Frosty to "get it." I didn't like where this was heading. I suspected that Frosty was in the process of reorganizing his understanding about what mixed numbers are, and even the clearest explanation about "where the 3 came from" might not be useful. I wondered if he was thinking about 3⅞ as two separate and unrelated objects: a whole number and a fraction. Not understanding 3⅞ as *a quantity* could help to explain why he hadn't been persuaded when Paul had added up from 3⅞ to 7 on the open number line (even though Frosty himself had used that same strategy to figure out 74 – 26 in the video clip). This conundrum underscores the limits of explaining: when students are in disequilibrium about mathematical concepts and relationships, explanations often don't help. That's because students like Troy, Paul, and Josie have already done the cognitive work of building their own understanding. Frosty still needs more opportunities to do so. For many years, the wisdom in the *Mathematics Model Curriculum Guide* (California State Department of Education 1988) has guided me in these kinds of situations:

> We must recognize that partially grasped ideas and periods of confusion are a natural part of the process of developing understanding. When a student does not reach the anticipated conclusion, we must resist giving an explanation and try to ask a question or pose a new problem that will give the student the opportunity to contemplate evidence not previously considered. (14)

But now time really was running out. I decided to leave the Frosty question hanging and make sure Irene had a chance to explain her answer. Irene, ironically, had done exactly what Frosty did. She said, "I took 4⅞, which is 7 minus 3⅞ . . ." but she had then changed 7⅞ to an improper fraction. Before she could finish explaining, though, the bell rang.

The Number Talk ends in an unexpected way.
http://sten.pub/dd11

While students were getting up to leave, I asked Irene to come up and show me her strategy. But while I was talking to her, and the videographers were packing up, a conversation erupted near the doorway. Josie and Xander hadn't been able to let go of Frosty's dilemma. We can't see Josie, who is off to the side, but we can hear her continuing to explain. Then we hear Xander making this argument as Irene wanders up to the group (and the screen goes black):

**Xander:** Frosty, I want you to ___ me on this, okay?
**Frosty:** All right.
**Xander:** Which one's more: 3 or 3⅞?
**Frosty:** 3⅞.
**Xander:** 3⅞ is more than 3, right?
**Frosty:** Definitely.
**Xander:** So you gotta take away more than 3 [from 7].
**Frosty:** I'm lost.

What students learn from us! This questioning sequence is a real-time example of "funneling" (see Chapter 4, "The Questions We Ask"), meant to lead Frosty through a logical chain of steps so he would realize that taking something greater than 3 away from 7 would have to be less than 4. Not surprisingly, Frosty says he's "lost," but Josie isn't about to let this go. She and Frosty go to the board as Paul wanders up. In what follows, the pressure on Frosty to understand Josie's fervent explanation is almost excruciating. But before long they high-five, smiling, and leave for their next class.

## Further Thoughts

There are no perfect Number Talks. They involve so many in-the-moment decisions that we never know what might have happened had we chosen different questions, different directions, different paths. But in spite of the different paths I might have taken, I felt invigorated afterwards. The class had come to life during the Number Talk in a way I couldn't have anticipated, and the energy in the room was palpable. While I was doing my imperfect best to manage the discourse, these students, in many small ways, had taken on some of that responsibility, too, making the Number Talk something of a joint production. The students themselves—not all of them, of course, but enough of them—shifted from compliance to active, interested participation as they became confounded by the differences in each other's thinking. This vested interest is part of what makes mathematics *matter*. Sense making makes mathematics personal, and when it's personal, it comes to life. And that's how Number Talks can really make a difference.

Now I wonder, *How did that happen?* and *How can we get more kids to do what Frosty did?* The class culture was comfortable and kind, which also helped me feel safe to actively get students involved. I asked them to rephrase another student's thinking; I encouraged them to direct their comments and questions directly to each other; I had them talk in small groups. At first, they humored me; gradually, though, they seemed genuinely engaged. But engagement is hard to assess, equated as it often is with who talks and how much. Listening, too, is a form of engagement, yet it's hard to tell by looking at the well-practiced blankness of high school students' faces to know who is engaged and who isn't.

In this Number Talk, nine students spoke to the whole class—some, of course, more than once—while others uttered not a peep. One young man can even be seen, head bobbing, as he tries (or not) to stay awake. Others, like Josie, became personally invested in the discussion. The discussions were not perfect, by any stretch of the imagination. Students had trouble explaining what they meant, and there were times when I could have supported more productive discourse by helping students make their explanations clearer to others. No resolution was ever reached, so by some standards, this may have been viewed as a failed lesson. One thing that matters to me, though, is that Xander, Josie, and, later, Paul cared enough to hang around after the bell rang, wanting Frosty to understand. This was a special moment, and I can't help but wish that I could spend more time with these kids, even for just a few more Number Talks.

In *Making Number Talks Matter*, we discuss the important and potentially necessary role that cognitive dissonance plays in mathematical learning (Humphreys and Parker 2015, 29–30). Number Talks can uncover, as we saw here, long-held misconceptions about numbers, operations, and mathematical properties that remain hidden as students move from grade to grade, only to emerge in unexpected ways that interfere with their success in later math classes. Getting a chance to *see* a student in the throes of disequilibrium is rare, though, and was all the more exceptional here because it was caught on video. But while it's tempting to conclude that Josie's picture of seven circles was all it took for Frosty to "get it," I'm not so sure. Despite the triumphant high-five, the way Frosty kept talking about the fraction being "over there" makes me think that his cognitive dissonance might not be so easily resolved by Josie's explanation—at least not for long. Something else I recently noticed in the video also makes me continue to wonder.

Frosty describes his strategy.
http://sten.pub/dd12

Frosty said, "What I did was, I took 7 and subtracted 3, and then *just put the fraction back on*. So it was $4\frac{7}{8}$." As I began to record, he repeated his method, with just one small variation: "7 minus 3 is 4; then I just put the $\frac{7}{8}$ back on." I was pretty sure I understood what he meant, so I thought I was revoicing his own idea when I said, "You put the $\frac{7}{8}$ back on, *so you added* $\frac{7}{8}$." But now I'm not so sure. Is putting something back on the same thing, for Frosty, as adding it? And I'm still wondering about what Frosty thinks $4\frac{7}{8}$ *means*. What, for example, might he do if I asked him to put $3\frac{7}{8}$ on a conventional number line? How else could we find out? It's fascinating, really.

Finally, I'm reminded of the constant lament we hear from high school teachers: "When are you going to write Number Talks with high school content?" I can't help but wonder, though, how this is not high school content. Admittedly, rational numbers are not in the high school curriculum, and moving abruptly from a subtraction Number Talk, for example, to an unrelated topic like solving systems of equations can feel inappropriately disconnected. And it's true that the fifteen Number Talk minutes are minutes you won't be able to spend on systems.

But Frosty wasn't the only one who got 4⅞; Irene, the last student to share, had also gotten 4⅞ before she changed it to an improper fraction. Even more disturbing to me is that the twenty-minute back-and-forth debate hadn't made Irene wonder about her answer. She didn't seem bothered at all. What does this say about how kids view math?

It's infuriating that so many students still come to Algebra 2 with such fragile understandings of quantities, operations, and numerical relationships. This is an indictment of a system that, for decades, ignored the growing body of research demonstrating that mathematics instruction in the United States was failing to adequately prepare its students for a changing world (Darling-Hammond and Adamson 2010; Hiebert 1999). Even as teachers are learning to implement the Common Core State Standards, they are faced with students, like Frosty, who have been failed by an archaic educational system where step-by-step rules take precedence over students' own reasoning. Even when their textbooks offer occasional nods to manipulatives and multiple representations, or when teachers try to explain why procedures work, rules are woefully inadequate for understanding big mathematical ideas.

How are students to make sense of mathematical content when, to many of them, mathematics is just a bunch of disembodied, personally meaningless procedures? How can students truly understand and apply rational numbers to variables and expressions, functions and equations *without* Number Talks? Yes, this was a Number Talk with high school content.

CHAPTER

# 2 Mistakes

Mistakes have gotten a bad rap in US mathematics classrooms. In a culture where quick right answers have been the measure of success in math, mistakes are regarded as deficiencies. Even the word *mistake* carries a whispered burden. And for too many of us, an element of shame creeps in.

Number Talks help reframe the role of mistakes. During Number Talks, students are invited to make sense, and teachers listen to their ideas instead of directing their thinking. This way, mistakes "provide opportunities to look at ideas that might not otherwise be considered" (Humphreys and Parker 2015, 27) and uncover misconceptions that otherwise might be hidden behind the veil of procedural accuracy. But even when teachers believe that mistakes are normal and necessary for learning, there is still a nagging responsibility that follows right behind: How will students learn if we don't point out their mistakes?

This chapter is prompted by the many questions we continue to receive about how to handle mistakes during Number Talks. We don't pretend to know a "best way" to handle mistakes as they arise, in their many variations; in fact, we're still tinkering with different approaches. Our thinking, though, is grounded in the belief that if mistakes are normalized, then space opens up for students to be curious enough about their own and others' mistakes to talk about them. And when students analyze their own thinking, the authority for correctness gradually shifts to their own reasoning and to the logic of mathematics.

Mistakes, and how they are handled, are central to whether Number Talks realize their potential. But when we began to dig into the role that Number Talks can play in shifting the class culture around mistakes, we uncovered many profound issues—too many to consider in this chapter. Therefore, we decided to share our current thinking, along with video examples, of what teachers can do in two particular situations during Number Talks: (1) there are wrong answers on the board at the end of a Number Talk, and (2) students make mistakes while explaining a strategy. Then we consider other common occurrences and offer ideas as you and your students renegotiate the culture around mistakes in your own classroom.

We emphasize, here, that these are ideas for you to try out and adapt to your unique setting and situation. Just as we support students to fiddle around with mathematical ideas, we hope you will tinker with these potential teaching strategies and see what works for you. And we hope these ideas will help you get started.

### When wrong answers are still on the board at the end of a Number Talk

One teacher expressed the dilemma this way: "I'm so uncomfortable leaving multiple answers on the board. I find myself wanting to have the class agree on the right answer. Help!"

In *Making Number Talks Matter* (Humphreys and Parker 2015), we recommend that teachers accept all answers and that they try not to indicate whether answers are right or wrong. But we don't talk about what to do with wrong answers that are already displayed on the board at the end of a Number Talk, and in our one-right-answer culture, this can be hard to bear. In the following video excerpt, Hailey is ending a Number Talk with third graders. Although four different answers are on the board, all of the students who shared their strategies supported the correct answer, which is 55.

At the end of her third-grade Number Talk, Hailey invites a defense of the other answers. http://sten.pub/dd13

When students don't defend other answers, Hailey moves on without comment. But what if Hailey instead, concerned that her students might be led astray into thinking that any of those answers could be right, had said, "So can we agree that the answer is 55?" Let's consider the implications, knowing that either (1) all students understand that 55 is the right answer, and why, or (2) they don't. In either case, there's a lot we don't know. We don't know if the students who initially gave wrong answers have changed their minds; we also don't know what students who didn't share are thinking. So, what to do?

First let's assume that all students indeed understand that 55 is the right answer—and why. It *is* possible that the students who offered alternative answers and those who didn't speak at all realized, by the end of the Number Talk, why 55 is the correct answer. We don't know this is so (but we think it's unlikely). Only Tiancum, having initially said "45," had publicly changed his mind by the time he shared a strategy. For the others, we just don't know.

Tiancum offers 45 but defends 55.
http://sten.pub/dd14

So, if all students have by now convinced themselves that 55 is the correct answer, then asking for a collective agreement would be unnecessary and might even be counterproductive. As Hiebert et al. (1997) point out,

> Part of what it means to understand mathematics is to understand the problem and the method[s] used to solve it. When this happens, the solver knows that the answer is correct. There is no need for the teacher to have the final word on correctness. The final word is provided by the logic of the subject and the students' explanations and justifications that are built into this logic. (40)

Hailey's endorsement of the answer would only serve, then, to shift the locus of authority from the students' own reasoning back to her own authority as arbiter of correctness.

Now, let's assume instead that most of the students have made sense of at least one of the strategies and understand why 55 is correct. But some students—we don't know which ones—might still be holding on to one of the other incorrect answers or something completely different.

We're skeptical, though, that asking the class to agree on one answer would benefit students who are still hanging on to other answers. Knowing that an answer is wrong—or even knowing why another answer is right—doesn't necessarily help us understand why our own answer is wrong.

No one in Hailey's Number Talk took her up on the invitation to talk about the other answers. But we've found that over time, as mistakes become a more natural part of the class culture, students begin to talk more freely and publicly about their own wrong answers, as Frosty does in Chapter 1. For these reasons we recommend leaving all answers, unsullied, on the board. The message to students? It's normal to have different answers. One is right, the others are wrong. No big deal! But students are not used to this strange practice of having teachers leave several answers on the board. They're used to having teachers be explicit about right answers. They're probably wondering, *Why doesn't she tell us which answer is right?* So, we need to let them know that we are doing this on purpose—and why.

**When students make mistakes as they are explaining the correct answer**

Students also make mistakes while explaining their strategies. Having space to resolve those errors on their own helps students learn to rely less on their teachers and more on themselves as sense makers. In the video that follows, we have the rare opportunity to watch a third grader, unprompted, resolve her own cognitive dissonance. The teacher, Hailey, has posed the problem 81 – 26, and Megan explains her strategy:

$$\begin{array}{r} 81 - 26 \\ \underline{-1 \quad -1} \\ 80 - 25 \end{array}$$

$$80 - 5 = 75$$
$$70 - 20 = 50, \text{ so } 75 - 20 = 55$$

But Hailey gets a bit behind in her recording and says, “Wait a minute,” as she tries to catch up. She checks with Megan: “You said 7 minus 20 is 50?” Although Megan had said “70,” Hailey had heard “7.” But Megan nods—and then says something different: “So 75 minus 25 is 55.” Hailey didn't seem to notice, but Megan did. Here, we watch as Megan resolves the mistake all by herself.

Megan figures out what happened.
http://sten.pub/dd15

Megan knows something isn't right—we can see it in her face and body language. But it's fascinating to watch as she whispers to herself, unable to attend to anything else around her until she has resolved her own disequilibrium. And we can *see* the moment when it happens; Megan's shoulders relax and she looks around. Only then can she tune in to what the next student is sharing.

In another example, this time from a middle school algebra class, Seth makes a couple of mistakes. His teacher, Nisha, gives him time to resolve them on his own.

Seth corrects his own mistakes.
http://sten.pub/dd16

It's hard to defy our natural inclination to jump in to correct errors, but Nisha barely pauses when Seth says 3 (instead of 8). A bit later, Nisha writes 328 (rather than 324, the correct answer) without batting an eye. We can be pretty sure Nisha is aware that those were mistakes, but it isn't evident in anything she says or does. This gives Seth the space to find and correct his own mistakes: his job, not hers.

A similar example (with no hint from the teacher that anything is amiss) comes from a fourth-grade Number Talk. Here Danni, as she is explaining her strategy, uses *minus* incorrectly.

Danni corrects her own mistake.
http://sten.pub/dd17

When Danni says, "7 minus 60 is . . . 7 minus 60 is . . . ," Hailey matter-of-factly records "7 – 60." It would have been easy for her to dismiss Danni's wording as a slip of the tongue and correct it. Instead, she records exactly what Danni says, without comment. Then Hailey, still without comment, draws a line through what she had first written when Danni changes her mind. Hailey, like Nisha, takes seriously a new teaching role during Number Talks: to simply elicit students' ideas and ask questions rather than pointing out their mistakes.

In these examples, Megan, Seth, and Danni were in the process of correct explanations that went only momentarily awry. But what if the students are defending flawed strategies or ideas?

## When students suggest flawed strategies—and realize something is wrong

In both of our experiences, students often realize that something is wrong, but they initially, at least, don't know what. Here are two examples, both involving the distributive property.

**1. Justin finds his mistake.**

Ruth posed the problem 26 × 48 to a fifth-grade class in Bellingham, Washington. (You might want to try to solve this problem mentally before reading on!) Justin was the first to volunteer, defending his answer, 848, this way:

$$\begin{array}{r} 48 \\ \times\ 26 \\ \hline \end{array}$$

$$20 \times 40 = 800$$
$$6 \times 8 = 48$$
$$800 + 48 = 848$$

Ruth knew that 848 was not the correct answer, but, without commenting, she left Justin's solution on the board, asking who thought about it differently. Here were some of the responses:

48 quarters = \$12.00 so $48 \times 25 = 1200$
$1200 + 48 = 1248$

$2 \times 48$ is 96, so $20 \times 48 = 960$
$5 \times 48 = 240$; $960 + 240 = 1200$
$1200 + 48 = 1248$

$26 \times 48 = 52 \times 24 = 104 \times 12$
$12 \times 100 = 1200$; $12 \times 4 = 48$
$1200 + 48 = 1248$

$50 \times 26$ is half of 2600, or 1300
$1300 - 50 = 1250$; $1250 - 2 = 1248$

$26 \times 48 = 52 \times 24$
$50 \times 24 = 1200$ (half of 2400)
$2 \times 24 = 48$; $1200 + 48 = 1248$

All of the solutions (except Justin's) supported the answer of 1248. Ruth waited a bit for other explanations and, when none was forthcoming, started to end the Number Talk. But Justin's hand shot up. "Wait!" he said. "I want a chance to disagree with myself!"

What happened with Justin occurs frequently in classrooms where students are comfortable making mistakes. Although Justin had been quite confident in his defense of 848, he soon became convinced, as he listened to the other explanations, that 1248 had to be the right answer. This led him back to his original strategy to think about what went wrong. He explained that even though he did $20 \times 40$ and $6 \times 8$, he forgot to do $20 \times 8$ and $6 \times 40$. Ruth wasn't surprised at Justin's mistake. She had used this problem many times before and had come to expect that 848 would, even with adults, be suggested. This is a conceptual error caused by some confusion about how numbers and properties work. Justin was convinced by the logic of others' strategies, and it was important that *he* was the one to find and correct his mistake—no one else.

In the next example, Kai makes a similar kind of mistake involving numbers and properties, but, unlike Justin, he doesn't find his own error.

**2. Kai doesn't know what went wrong.**

The second example comes from high school geometry, where the students were just starting out with Number Talks. Tamara, their teacher, had been trying to help them break free of the standard algorithm for multiplication by choosing problems that beg to be rounded. Towards the end of their previous Number Talk ($18 \times 11$) Tamara had suggested a "strategy I saw in Ms. Yu's class." She showed them how a student had rounded 18 to 20, thus changing the problem into the easier $20 \times 11$; then the student only needed to subtract 22 from 220. She left the idea hanging and moved on to the day's regular lesson.

Two days later (they did Number Talks twice a week), Tamara posed the problem $12 \times 28$. Only one answer (336) was offered. Students had several strategies: $(10 + 2) \times 28$; $3 \times 4 \times 28$; $12 \times 2 \times 14$. But no one rounded. Finally, Kai raised his hand. "I got a different answer," he said, "so I know something is wrong, but I don't know what." (We've found that when the right answer is the first one to be shared during Number Talks, students—especially in middle and high school—can be reluctant to share an alternative answer. This is normal when Number Talks are just getting started.)

**Tamara:** What answer did you get?

**Kai:** 304. I tried that rounding strategy you showed us. I rounded 28 to 30.

**Tamara:** So you multiplied . . .

**Kai:** 12 × 30.

(*Tamara recorded 12 × 30 and waited.*)

**Kai:** I multiplied 12 × 30 and got 360, and then I subtracted 56 and got 304.

(*Tamara recorded 12 × 30 = 360*
*360 − 56 = 304*)

**Tamara:** Where did the 56 come from?

**Kai:** 2 × 28 = 56. I subtracted two 28s because I rounded.

This is a tricky and critical point in this Number Talk. What are some things Tamara could do?

**1. She could explain how to correctly compensate for rounding in multiplication.**
Tamara's first inclination might be to step in and show Kai how to do the problem correctly. After all, that's what most of us were taught to do! She might demonstrate something like this:

$$12 \times 28 = 12 \times (30 - 2) = 12(30) - 12(2) = 360 - 24$$

The trouble is, this decision would be based on the (mistaken) belief that a procedural explanation can remedy a conceptual issue. Kai's misunderstanding about how numbers and their properties work is actually quite fascinating and worthy of further exploration. More troubling, though, is that showing Kai what to do would immediately shift the authority for correctness right back to the teacher. And since Number Talks are about helping students *own* their ideas and understandings, Tamara didn't make this choice.

**2. She could turn Kai's question back to the class.**
Deciding to do this, though, depends a lot on the existing class culture. How do students feel about being wrong? What do they think it means to be good at math? Have they become genuinely interested in one another's ideas, beyond the rightness or wrongness of answers? Are they comfortable enough to risk sharing their own mathematical ideas when perhaps they aren't sure?

We are not talking, though, about a community effort to "help Kai," which violates the intent of Number Talks. What if Kai doesn't want to be "helped"? We're talking, instead, about a collective willingness to consider mathematical ideas together. The decision to have everybody engage around Kai's idea depends not only on Kai's willingness to subject his thinking to the scrutiny of his peers but also on how well the class has learned to engage in collaborative inquiry. This takes time. At the time, Tamara didn't feel that her students were quite ready to engage in these ways, so she opted instead for the third choice: save the question for another day. She thanked Kai for sharing his question and promised the students that they would return to the idea in another Number Talk.

**3. Tuck the idea away and use it (soon) for a short experience in investigative problem solving.** In situations like the one Tamara faced, we hold on to one of our Guiding Principles for Number Talks (Humphreys and Parker 2015, 29): mathematical understandings develop over time. When students reveal "soft spots" in their understandings, teachers have an opportunity to pose another problem that might help students look again. So, when an interesting puzzle arises like Kai's, we can use the next Number Talk time to take a deeper look. To reengage with the underlying relationships in Kai's problem, we might create a new problem like the one that follows. We wanted one factor to "call out" for rounding up, and we considered using 18 or 19, but decided that 18 (or 28, 38, etc.) would keep the cognitive demand higher. Here is a problem we might pose:

Sanjay and Selena had just done this Number Talk in their class: 18 × 32.

They both started out the same way (by rounding 18 to 20), but ended up with two different answers:

| Sanjay's strategy: 20 × 32 = 640 | Selena's strategy: 20 × 32 = 640 |
|---|---|
| 640 − 36 = 624 | 640 − 64 = 576 |

Which *strategy* do you agree with? Why? Your job is to analyze the strategy. (Caution: Finding the answer a different way and using that answer to defend your choice is not proof enough!)

First, convince yourself. Then, using words and/or diagrams, convince a skeptic.

With a lesson like this, we would first have students work independently for a while to have time to gather their own ideas. Then we would have them get together in small groups to share their thinking. Once students are prepared to convince others why one of the solutions makes sense—or not—we would bring the class back together for whole-group processing. We would ask if any group would be willing to get the conversation started by sharing their findings. As in a regular Number Talk, we would continue to ask if other groups have different convincing arguments. Posing a problem in this way can help all students dig in, and it can deepen their understanding of the mathematical relationships involved.

There are, of course, other ways to reengage the class around an idea that has arisen during a Number Talk. For more examples, see *Making Number Talks Matter* (Humphreys and Parker 2015, 133–162).

## When students suggest flawed strategies but don't know they are flawed

Sometimes, students share an idea that doesn't make any sense. In the video clip that follows, for example, Mica is working on the problem 62 – 29. It took a while to figure out what she was doing, but it looks like she is using an add-up strategy for subtraction that she really doesn't understand. With this strategy, students start with the subtrahend (29) and add one number at a time to reach the minuend (62). Then to find the answer, students circle the numbers that were added. If Mica had used the strategy correctly, she would have gotten some variation of this (see Figure 2.1):

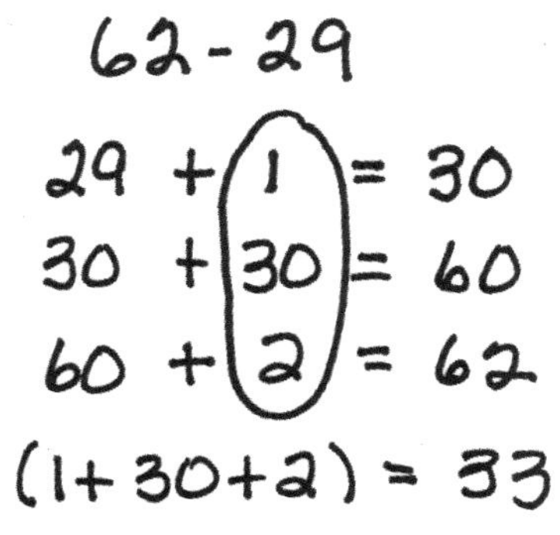

62 - 29
29 + 1 = 30
30 + 30 = 60
60 + 2 = 62
(1 + 30 + 2) = 33

**FIGURE 2.1**

Mica, however, started with the minuend and *added* the subtrahend (see Figure 2.2).

$62 - 29$

$60 + 20 = 80$

$80 + 2 = 82$

$82 + 8 = 90$

$90 + 1 = 91$

$(20 + 2 + 8 + 1) = 31$

**FIGURE 2.2**

Mica misapplies a subtraction strategy by adding the numbers.
http://sten.pub/dd18

We see that Hailey accepts Mica's strategy with no comment. This situation was best left unexamined.

## A student wants to talk about someone else's mistake

When a student makes a mistake during a Number Talk, it's natural for others to want to point it out. But while it's one thing for someone to talk about their *own* mistake (as Frosty did in Chapter 1), it's something else entirely for someone else to talk about it. We do want students to learn to examine and analyze others' thinking, over time. After all, "critiquing the reasoning of others" is one of the mathematical practices that we hope to nurture (NGA/CCSSO, 2010).

Having the class focus on one student's wrong answer or faulty logic, though, has the potential for unwelcome, anxiety-provoking scrutiny. And if the intent is to "fix" or even "help," then we don't do it. Number Talks are not about fixing students' thinking—and there's no difference whether the help comes from a student or from the teacher. When a student wants to comment on another student's mistake, therefore, we make sure to ask the student if he is open to hearing others' ideas or whether he would rather have more time to think about it on his own. This way, we avoid the common trap of getting the class to help fix someone else's mistake.

Mistakes, confusion, and misconceptions that might usually remain hidden are regularly unearthed during Number Talks. We see these as opportunities. The traditional model of correcting errors hasn't served our students very well, but here are some more ideas for addressing students' attitudes about mistakes without interfering with their own sense making.

## Normalize Mistakes

Probably one of the biggest challenges for teachers is to cultivate a learning community where students take their mistakes in stride. We offer, here, our ideas about how to make mistakes a normal, natural, and useful part of learning.

### Dedicate time in class for students to reflect on and share their own mistakes.

Asking students to reflect on their mistakes and missteps can help them recognize that mistakes are a natural part of learning and nothing to be ashamed of. Near the end of a Number Talk, for example, we could ask questions like these:

- Can anyone think of a mistake you made that you would be willing to share?
- Did anyone choose a strategy that didn't work very well?
- Is anyone willing to talk about where your thinking got bogged down?

### Share brain research.

Shifting the role of mistakes in a math classroom, then, means relearning with each other—teachers and students alike—how to think and feel about our own and others' mistakes. Many teachers find it valuable to share with their students research that shows how mistakes grow our brains (see, e.g., https://www.youcubed.org/?s=mistakes).

### Share your own mistakes.

Our own mistakes are opportunities too, and how we handle them can make a big difference in how students view their mistakes. If we are embarrassed by our mistakes and try to cover them up, or even joke that we made the mistake "on purpose," we reinforce the idea that mistakes are not okay. But when we openly acknowledge our mistakes, take them seriously, and even express interest in them, we send a different message. Everyone makes mistakes during Number Talks; they are part and parcel of the everyday learning of mathematics.

Here's an example of a mistake Cathy made as she was working with elementary teachers in Stanford University's Teacher Education Program. The video has been edited but includes enough of the

Number Talk to give you context for what happens when she shifts the focus from solutions for 25 x 29 to the properties of real numbers.

Cathy uses a multiplication Number Talk to examine the distributive property.
http://sten.pub/dd19

Joe was the first student to share an answer (720), but when Shea offered her answer (725), Joe immediately wanted to change his. He explained that he had first changed the problem to 25 × 30, had multiplied 25 × 30, and that's when he got confused (which, incidentally, was the same place in the problem that Kai got mixed up in the high school Number Talk earlier in the chapter; it's tricky to figure out how to compensate after rounding up in multiplication). Unlike Kai, though, Joe figured out his mistake. He explained that when he multiplied 25 by 30, he added an extra 25 and should have subtracted 25 instead of 30.

Then I drew a diagram that represented how Joe solved the problem and asked students to talk about what property they thought he had used and how they would represent it algebraically. After they talked awhile, Kellen volunteered, identifying the diagram as a representation of the distributive property, and said the algebraic expression would be 25(29 + 1), which I recorded. It wasn't until later, though, while viewing the video, that I realized that 25(29 + 1) didn't actually represent Joe's thinking. Joe had multiplied 25 by 30 and then subtracted 25, so the correct representation would have been 25(30 – 1). I didn't even notice.

When I finally realized my mistake, I hesitated about including the video in this book. It's one thing to talk about sharing our mistakes openly, but another thing altogether to actually do it! ☺ As Lampert (2001) points out, however, it builds character:

> One of the hardest things to do in front of a group of one's peers is to make a mistake, admit it, and correct it. Yet such a series of actions is an essential component of academic character. It defines the very nature of learning. (266)

But, people ask, what *would* I have done had I recognized the mistake right away? Hmm. In hindsight, I think it would have been good to just leave Kellen's mistake on the board without commenting and then ask if anyone had thought about it differently or found a different expression. If no one did, I definitely would have tucked this away for another day.

We can help students more easily make friends with their mistakes and embrace them as opportunities to learn if we, ourselves, get better at welcoming our own mistakes.

## Be transparent about your own learning.

Being transparent about our own learning can also help students believe that our messages about mistakes aren't merely platitudes. Gloria Moretti, our friend and colleague, spent several years in a teacher leadership project that required frequent absences from school. She made it a habit, when she returned, to talk with her students about her experiences. She says,

> It was usual for me to share what I was learning, talk about why we were doing things differently, and highlight ideas for them to consider, like how mistakes are an important part of learning. And I think this practice made an unanticipated contribution to how we developed as a caring community. It seemed to me that students related to and accepted the idea that when you are learning something new, you can expect to make mistakes.

## Put the focus on figuring out.

Number Talks can provide the grist for reasons to dig in and figure things out—to find out *What's wrong, and why?* When we focus on figuring things out, then we stimulate the natural human desire to want to know why things work—or why they don't. It usually doesn't take long, then, for students to begin saying things like, "I know my answer is wrong, but can I share what I did?" or "I tried something but I don't know why it didn't work," or "I've been thinking about what Keith said, and I was thinking . . ."

Johnston (2004) makes a compelling case for letting students make mistakes and struggle to figure things out on their own. He suggests that if we explicitly and repeatedly tell students what to do and how to do it, they start to believe that they aren't the kind of people who can figure things out for themselves. He writes, "Think about it. When you figure something out for yourself, there is a certain thrill in the figuring. After a few successful experiences, you might start to think that figuring things out is something you can actually do" (8). And, although it takes time and patience to shift students' dispositions about mistakes, we can be buoyed by the work of Alan Schoenfeld, who found that when students are allowed to "mess around and live with some chaos," they fix their own mistakes *nearly all of the time*. He also cautioned about the dangers of intervening in ways that "step on students' sense making" (2017; italics added) and prevent mistakes from happening.

It's a challenge to reframe how mistakes are viewed in math class. But Number Talks can help. With Number Talks, students learn to expect math to make sense. And when students expect things to make sense, they are more likely to summon the courage (and even risk their social status) to share an answer even when they suspect it might be wrong, or to say things like, "I got a different answer, but I think I might know what I did wrong," or, "I got this far, but then I got stuck." These are small indicators of a big shift within the hearts and minds of students who are starting to believe in their right to understand.

CHAPTER

# 3 Revising the Number Talk Routine: Rethinking Wait Time

> ***To "grow" a complex thought system requires a great deal of shared experience and conversation. It is in talking about what we have done and observed, and in arguing about what to make of our experiences, that ideas multiply, become refined, and finally produce new questions and further explorations.***
>
> —MARY BUDD ROWE, "WAIT TIME: SLOWING DOWN MAY BE A WAY OF SPEEDING UP"

Many teachers strive for the kind of classroom dynamic Mary Budd Rowe describes, but find it hard to achieve. Even as more and more math classrooms move toward collaboration and problem solving, it can be difficult for teachers to generate and sustain an atmosphere of student-to-student interaction, particularly in whole-class discussions. In this chapter, we look at how wait time can be used effectively in Number Talks to ensure more equitable participation, bring more voices into the conversation, and help students learn to engage with each other's ideas.

Over the past couple of years, as we've watched and reflected on Number Talks, we've noticed that student-to-student interactions are often lacking. As teachers ask students to explain their methods for solving problems, record their thinking, and probe for why those methods make sense, it can be easy to fall into a "teacher-student-teacher-student" pattern of discourse that leaves little space for spontaneous student comments or conversations. In the past, we've tried a variety of ways to encourage students to engage around others' thinking—but with mixed results. After a student has explained his strategy, for example, we've experimented with questions to help students see that there is a place for their ideas in other students' thinking. But questions such as "Who has a question for _____?" or "Who has a question that would help us understand _____'s thinking?" sometimes result in superficial responses like "Why did you use that strategy?" or even utter silence. As we pondered our discomfort with what we came to view as "flatness" in some Number Talks, we became convinced that we need to purposefully build more opportunities into the Number Talk routine for students to reflect on each other's ideas and provide more space to nurture their curiosity and wonderings. But how?

Talking this over (and watching a lot of videos) brought Mary Budd Rowe's research on wait time (1986) to mind. She found that two different places to wait—after the teacher asks a question and again after a student's response—"make a fundamental impact on the reasoning, roles, and norms in a classroom" (43). The first wait time has been important in our teaching for years, but we had both forgotten how important it is to wait a bit after a student responds before saying anything ourselves. Rowe found that when teachers extend both of these wait times, students learn more. They also ask more questions, the number of student-to-student exchanges grows, and the "variety of students participating voluntarily in discussions" increases (44). This made us think that being more intentional about wait time will potentially shift the classroom dynamic so that the mathematical discourse becomes more authentic, more interactive, and more exploratory.

Rowe's first wait time is a natural part of Number Talks. When teachers pose a problem and wait until most thumbs are up, students have plenty of time to think about the problem. But in the past we haven't done justice to this first wait time when we ask, "Who is willing to share their strategy?" We've watched many videos of Number Talks—including our own—and have noticed how quickly teachers call on students to share. It takes time—certainly more than a few seconds—for students to gather their thoughts or muster the courage to share their thinking. And if we habitually call on those who are first to raise their hands, it's likely that those same few students will continue to do most of the sharing. Our silence *after* asking for strategies but *before* calling on anyone can be hard for students. They may not understand why we're just waiting when their hands are up or why, if we catch their eye, we don't immediately call on them. As with other culture-shifting decisions we make during Number Talks, we need to let students in on our thinking about *why* we are waiting and *why* it's important to give everyone time to think about whether and what they would like to share. Waiting is new to us as teachers, too, and hard to remember, as Ruth shows us in the vignette that follows. But waiting before calling on anyone shows such promise as a way to encourage more equitable participation during Number Talks. So providing all students the time they need to think and decide on their own if they're ready to share with the class is time well spent.

Rowe also found that the increased student-to-student exchanges, the very thing we've been longing to see more of in Number Talks, were particularly influenced by the second type of waiting: after a student has responded. We're now determined to be more purposeful about waiting for several seconds after a student is finished describing her strategy before we respond in any way. Providing this space gives other students a chance to process what has been shared and shifts their responsibility from just listening quietly (until they can share their own strategy) to also actively considering the thinking of others. We envision these moments as the time during Number Talks when students can gradually learn to comment spontaneously on each other's ideas without raising their hands.

Finally, to build on Rowe's ideas, we think there is yet another place where waiting can make Number Talks even more powerful and purposeful: when a Number Talk is ending. It can feel a bit unsatisfying to simply thank students for their participation, so we have been tinkering with what to do about this. The end of a Number Talk, when all the strategies are visible, is full of potential. Students can look across the different strategies, take stock, look for connections among the various strategies, digest new ideas, and think about new questions or wonderings. As we talked about this, we realized that if we want students to engage in this kind of thinking and analysis, then we have to dedicate space for it to happen but *without* directing their thinking in any particular direction. What do *they* notice? What are *they* wondering about? New wonderings and "soft spots" in students' understandings will emerge during this final wait time. And when it's not possible to deal with them in the moment, it can help to jot down the ideas and plan to pursue them in future Number Talks, build them into upcoming lessons, or use them to launch an investigation. The following is a vignette from Ruth, "Giving It a Try: Change Is Hard."

### *Giving It a Try: Change Is Hard (A Vignette from Ruth)*

Not long after making this wait time commitment to myself, I had an opportunity to visit a school for girls in San Francisco where some teachers had been working on Number Talks for a few years. I was asked to do Number Talks in second- through eighth-grade classrooms, with teachers watching. I mentioned to Karen, the school leader who would accompany me into classrooms, that I was excited to practice some new ideas about wait time during Number Talks. But in spite of my excitement, during the first two Number Talks I did that day I forgot to wait! Finally, in my visit to a fourth-grade classroom, I remembered. Here's what happened.

Before writing a problem on the chart paper, I told students that they would probably just "know" the answer to the problem, but that that was not what I was interested in today. I asked them to think about what they could do if they didn't already know the answer to the problem. How could they use what they already know to make this an easy problem to solve? I told them that I was going to give everyone time to think about this and that they should let me know with their thumb when they found a way, or more than one way, to make the problem easy (see Chapter 5: "Nudging").

I then wrote 12 × 9 on the board and waited as students indicated with their thumbs that they had a way to find the answer (most students indicated at least two ways). I then asked if anyone was willing to get us started by sharing their answer. I continued to ask for different answers and recorded the three that students shared: 100, 108, and 118. After getting multiple answers, I commented, "Great. Now we really have something to talk about because we have different answers," and asked if anyone was willing to convince us their answer made sense by telling us what they had done. And I remembered to wait.

Before anyone volunteered, Tamika said she didn't think it was 100 anymore. I crossed 100 off the list and asked if she wanted to say more about this, but she shook her head.

Jessi volunteered that she had counted by twelves, and I recorded as she said, "12, 24, 36, 48, 60, 72, 84, 96, 108," quite confidently. When I asked her how she knew where to stop, she shrugged her shoulders and said, "It just felt right." I remembered that third wait time, looked at what I had written, and began silently counting to ten. Alexis jumped up, saying, "There are nine twelves up there—look!" and going to the chart, she counted the nine twelves, at which point Jessi said, "Oh yeah, I kept track on my fingers till I had nine twelves."

I asked who had another way that made the problem easy to think about. Alex said she did 10 × 9 and that was 90. Then she added 9 and 9 and got 18 and then added 18 to 90 and got 108. When I asked why she added 9 and 9, she replied that she needed two more 9s because she only did 10 x 9. Again, I forgot to wait to give other students time to think about what Alex had said and immediately asked for additional ways to think about the problem.

Jamie said she did 6 × 9 because she knew that was 54, and then she added 54 and 54 and got 108. I asked how she added the 54 and 54, and she replied that she added 50 and 50 and got 100, then added the two 4s. I waited a few seconds in order to give others a chance to think about what Jamie had said.

I invited others to share, and Tamika asked if she could share what she had done wrong. I said, "Sure!" and then listened. Tamika said she had thought about it like Jamie, but then just doubled the 50 and forgot about the 4s, so she agreed now with 108.

Tamara went next, saying that she did 9 × 11 because "11s are easy." Then she added one more 9 and got 108. Alex was next, mentioning that hers was like Jamie's, but she thought about the 12 as 6 x 2, not 6 + 6 so she did 6 × 9 and got 54 and then doubled the 54 for 108. I was pleased that she had seen the somewhat subtle distinction between her method and Jamie's.

Our time was just about up when I *finally* remembered to try the fourth wait time. I told students that before we ended the Number Talk, I would like them to take a look at all the strategies we had on the chart paper and think about whether there was anything they noticed or anything they wondered about. And I'm so glad I finally remembered!

One hand went up right away, and I asked students to show me with their thumb instead, so that everyone would have a chance to think about what they had noticed or wondered. I waited until a few thumbs were up (about twenty seconds) and then asked if anyone would like to share something they noticed or wondered.

Maria said she noticed that we had three different answers for 12 x 9, and they were all even. She wondered why we didn't get any odd numbers for answers. Jaya noticed that some of the strategies were like each other but different, too. I asked if she would tell us more about what she was noticing, and she said, "Well, like, Alex did 10 × 9 and she had to add two more 9s, and Toby did 11 × 9 and she only added one more 9. So, like, when you use different numbers of 9 at first, you have to add different numbers of 9 at the end." I asked if others saw what Jaya was noticing, and many nodded.

Maria raised her hand again and said that she was still thinking about why all of the answers were even numbers. She thought maybe it was because the first number in 12 x 9 is even. But then she realized that couldn't be it, because 9 × 12 is the same as 12 x 9. I told the class that Maria was getting at a big mathematical idea—when would an answer be even, and when would it be odd?—but that we didn't have time to investigate it right then.

Our time was up, and I ended the Number Talk by thanking everyone for sharing their thinking. I told them I noticed how interested they had been in each other's ideas, how confident they were in sharing their thinking, and how willing they were to change their mind and talk about their mistakes. I let them know that those are all really important mathematical dispositions.

As I was leaving the classroom, the teacher stopped me to tell me that Maria came up to her and said she had a theory but had to go home to investigate it further. Maria's theory was that you would get an answer that's odd only when you multiply two odd numbers. She said that she tried it with 3 × 5 and 7 × 9 but that wasn't enough to be convinced, so she was going to think about it more at home that night.

Maria's theory brought to mind something I remember the mathematician Peter Hilton saying many years ago at a math conference in Southern California. Here's his message as I remember it. "In this country you think that mathematics is about going from questions to answers. But that's a fallacy; just a fallacy. What mathematicians do is go from answers to new questions." At the time, I remember thinking, *Ah, that's what we're trying to do: learn to teach in ways so that when kids have solved a problem, they don't think they're done. Instead, they take time to think about what they notice or what they're curious about now.* And it was happening in my very first time of practicing the fourth wait time! Maria would not have had time to even consider the issue of odd and even answers, let alone express what she had noticed, had I (once again) forgotten to wait. My experience in this fourth-grade classroom, and in other classrooms that followed, made me even more convinced that these four wait times during Number Talks are essential. They have the potential to nurture both curiosity and reflection in students—dispositions we hope every student will share. We find an example of this in the audio clip that follows.

I was surprised that it took me three Number Talks before I remembered to try these additional wait times. Habits of teaching are hard to break—not because we don't want to change, but because it is so easy to slip back into what's comfortable. Since using these extra wait times, though, I've become convinced that waiting during Number Talks can open up a whole new mathematical world for students.

In this audio clip, Jamie Souhrada, a teacher in Washington State, shares what happened in her eighth-grade math class the very first time she tried waiting at the end of a Number Talk to give students time to notice and wonder (see Figure 3.1; webinar, Mathematics Education Collaborative, January 2–18, 2018).

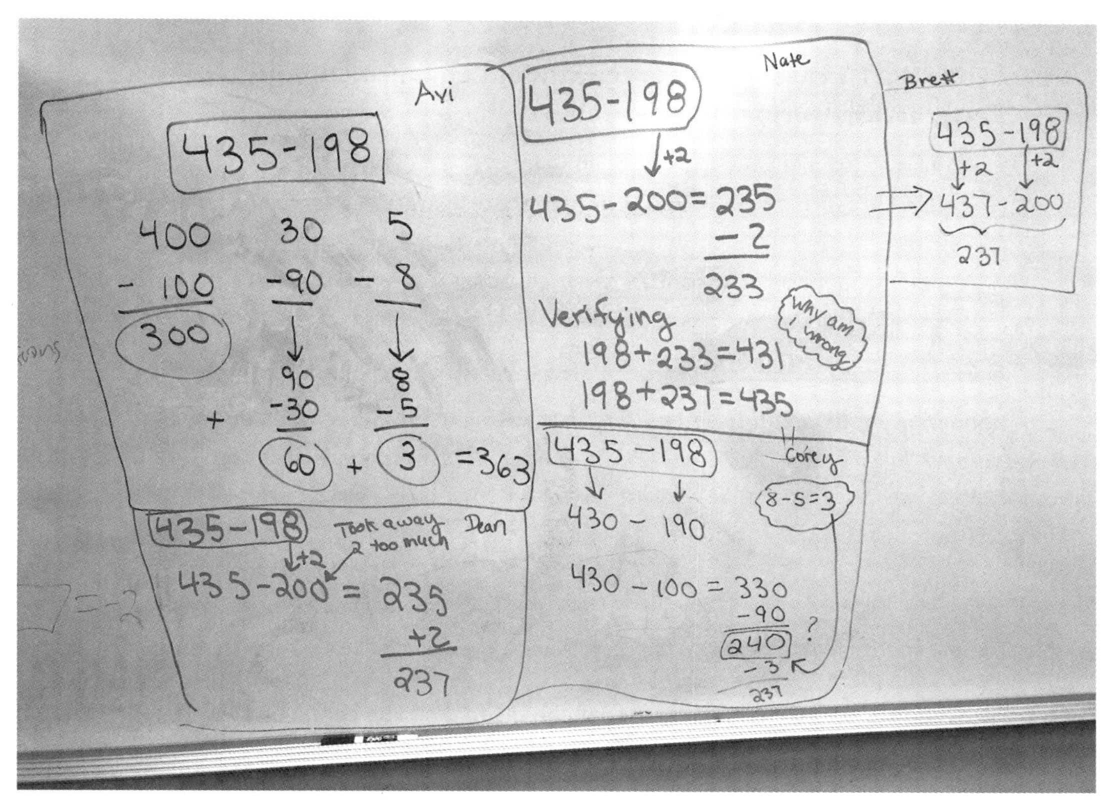


**FIGURE 3.1**
Strategies from Jamie's Number Talks.

Jamie asks students what they notice or wonder.
http://sten.pub/dd20

Students need time to make sense of mathematical ideas in their own ways and to discover mathematical connections in relationships between seemingly unrelated ideas. And they need time to reflect on what they are learning and what they are wondering along the way. Making good use of these four different wait times can go a long way towards helping make Number Talks more vibrant and interactive.

**Four Wait Times to Encourage Student Engagement and Interaction During Number Talks**

- *After* the teacher poses the problem
- *After* the teacher asks who is willing to share a strategy and *before* calling on anyone
- *After* a student responds and *before* the teacher responds
- At the very end of the Number Talk

Our pondering about wait time led us to refine the original Number Talks routine that appears in *Making Number Talks Matter* (2015, 11–13). This revised routine incorporates each wait time and its rationale, along with suggestions for what to say to students.

Prior to this routine, you'll want to try the following:

1. Dot Talks (pp. 65–74)
2. The first approach to nudging: "What could you do if you didn't already know the answer?" (p. 61)
3. The second approach to nudging: "What could you do if you hadn't been shown how to do this?" (pp. 61–62)

## Revised Number Talk Routine

1. **Ask students to put their paper and pencils away and put their closed fists on their chests to show you that they're ready.**

   Why: If their closed fists are already on their chests, it's easy for students to unobtrusively put up their thumbs when they've had enough time to think about the problem. Putting paper and pencil away tells students they will be solving problems mentally.

   Say (something like), *"Okay, please put your paper and pencil away and get ready for our Number Talk."*

2. **Write a problem on the board or document camera (generally, we write the problem horizontally).**

   Why: We write problems horizontally so students will be less likely to use standard algorithms.

   Say (occasionally remind students by saying something like), *"I'm writing the problem horizontally to remind you that during Number Talks, we don't use the standard algorithm."*

3. **Have students solve the problem mentally and put up their thumbs when they have had enough time to think; then WAIT.**

   Why: The purpose of intentional waiting is to shift the emphasis from being quick to being thoughtful. This is the first of four different places to wait during a Number Talk.

   Say (something like), *"Please don't raise your hands when you've solved the problem. It's not a race! I'm not interested in how fast you can solve it. I'm interested in hearing* how *you solve it. And besides, raised hands can interfere with people's thinking.*

   *"Putting your thumb up doesn't mean you have to talk. It just means that you've had enough time to think about the problem. We won't call on you unless you volunteer.*

   *"If you have extra time, see if you can find other ways to solve the problem and let me know by . . ."* (Model putting up additional fingers.)

4. **When most thumbs are up, let students know that you are going to collect answers, but that it's important that they not let anyone know whether or not they agree with an answer.**

   Why: Number Talks are designed to help students learn to reason with numbers. Any attention given to who agrees with an answer can detract from the essential focus of sense making. Also, some students will be less likely to share their answers if another answer seems more popular or prevalent.

   Say (something like), *"It's really important that you not let anyone know whether you agree or disagree with someone else's answer. We want everyone to share whatever answers they got. You'll have a chance to convince others that your answer makes sense if you want to, but we want to make sure that everyone feels safe to say their answer, whatever it might be."*

5. **Ask students to put their thumb up if they're willing to share their answer.**

   Why thumbs instead of raised hands: This is new and designed to address a common problem teachers have. All too often, the same few students raise their hands and it can be hard not to call on one of the first few hands that go up. Also, raised thumbs don't draw such noticeable attention to those who want to share, so there is less pressure to think fast.

   Say (something like), *"Okay, it looks like we're ready to get started. Put your thumb up if you're willing to share your answer."*

6. **Record just the answers on the board (noncommittally). Then continue to ask if anyone got a different answer that they are willing to share. Be sure to record each answer only once.**

   Why: When students begin to share their answers, don't indicate in any way whether an answer is right or wrong. This can discourage students from sharing alternative answers. Besides, Number Talks are more engaging and interesting when there is more than one answer.

   If only one answer is given, say (something like), *"What's going to start happening is that we'll get different answers for a problem. This is a natural part of Number Talks, so if you ever have a different answer, we hope you will share it."*

   If more than one answer is given, say (something like), *"This is good! We were hoping this would happen, because now we really have something to talk about."*

   In either case, say (something like), *"Sometimes, even at the end of the Number Talk we might still have different answers. One is right, the others aren't. No big deal. We don't need to agree on one answer, because the idea of Number Talks is for you to learn to use mathematics to convince yourself and others that your answer makes sense."*

7. **When there are no other answers, ask students to put their thumb up again when they are willing to explain how they figured out the problem. Then WAIT.**

   Why: This is also new. Some students love to volunteer; others are more reticent. It can take time for students to decide if they are ready to share and what they might say. Waiting here helps to avoid the problem of hearing from the same few students. And once again, thumbs are less intrusive than raised hands.

Say (something like), *"You may notice that sometimes I see your thumb and even make eye contact with you but don't call on you. This might feel like I'm not interested in what you have to say, but I am. I'm waiting on purpose to give everyone a chance to think about whether they want to share and what they might say when they do. Not everybody who wants to will get a chance to share in any one Number Talk. Over time, it's my job to make sure that everyone who wants to share will get a chance."*

8. **When volunteers begin to share their strategies, they first identify which answer they are defending. Record the student's thinking. Then, before calling on another student, WAIT.**

   Why: It's too easy to go from one strategy to the next without giving students time to think about a strategy that has just been shared. Without this wait time, we are often the only ones to comment or ask questions. These five to ten seconds of waiting can open up space for students to consider others' ideas. This also gives *you* time to think about what a student has done.

   Say (something like), *"You'll notice that sometimes I wait awhile after someone has shared a strategy. I'm doing this on purpose to give you a chance to think about what that person has said."*

9. **Ask if anyone used a different strategy that they would be willing to share. Then WAIT. Continue until no other thumbs are up (as time permits).**

10. **Ask students to look at the different strategies shared and see if there is anything they notice or anything they wonder about and indicate with their thumbs if they do. Then WAIT until several thumbs are up. And then wait some more.**

    Why: This is the (very important) time for students to look across strategies and notice relationships. But try hard not to do the thinking for them (no matter how tempting it might be, or how excited you are about something you've noticed).

    Say (something like), *"We have several strategies up here. Take a look and see if there are things that you notice or are wondering about. Let's give everyone time to think about this and you can let me know with your thumbs if there are things you notice that you would like to talk about."* And **wait**.

## Ending the Number Talk

Number Talks don't naturally end after fifteen minutes; often, they can go much longer if you let them—and sometimes you might want to let them. Either way, it's useful to think ahead of time about what you will say to end the Number Talk. It can be as brief as saying something like, "Thanks, everyone. I loved hearing your ideas and hope you also liked hearing the different ways people thought about this problem." Or, "Oh my gosh! Our time is up. We didn't get to hear from everyone who wanted to share, and I hope we get to hear from more of you tomorrow."

See Appendix C, "Revised Number Talk Routine in a Nutshell," for a quick reference guide.

CHAPTER

# 4 The Questions We Ask: What? Why? When?

Our questions matter. As reflections of our beliefs about the nature of mathematical learning, the questions we ask help shape students' ideas about what mathematics is and what it means to learn mathematics. In *Making Number Talks Matter*, we describe the principle that guides our teaching decisions during Number Talks: "Through our questions, we seek to understand students' thinking" (Humphreys and Parker 2015, 26). To us, "seeking to understand" means that we are genuinely curious about how students are thinking. We try our best to draw out students' understandings and help them put their ideas into words by asking questions like, "How did you think about it?" or "Why did you choose to ________?" Questions like these, without prespecified or expected answers, are often described as "authentic." According to Nystrand and colleagues (2003), "Authentic questions posed by the teacher signal to students that the teacher is interested in what they think and know" (145). Authentic questions are also genuine in the sense that we don't know the answers.

Not many of us, though, were asked authentic questions when we were math students. Instead, questions like ,"How many degrees are in the sum of the interior angles of a pentagon?" or, "What is the greatest common factor of 24 and 36?" have dominated mathematics classrooms for so many years that they are nestled snugly in our teaching DNA. These "known-information" questions (also sometimes called "test" questions) come out easily and naturally because of our experience as students, and usually the answers are either right or wrong. With these kinds of questions, it's easy to know how to respond. Authentic questions are harder. We can't know how a student will answer a question like, "How did you think about it?" or anticipate what we should do once they tell us. As one teacher commented, "It's so hard to ask questions! So hard!" Asking questions about students' thinking, therefore, remains one of the biggest challenges teachers face during Number Talks.

In this chapter, we dig more deeply into the hows and whys of teacher questioning during Number Talks, even while knowing that we can only scratch the surface of the issues that arise in classrooms. We discuss these issues:

- When and why to ask questions
- How to get students actively engaged with each other's thinking
- What kinds of questions can get in the way of student sense making and engagement
- What to think about as you ask questions

Finally, we offer general suggestions to make your Number Talks more successful. In Appendix A, we also list variations on questions.

## Eliciting Student Thinking During Number Talks

As we mentioned in *Making Number Talks Matter* (Humphreys and Parker 2015), there isn't a script for questioning during Number Talks. The only truly predictable questions come at the very beginning, when we invite students to share their answers or strategies. After that, our questions need to follow students' thinking. "Following" means that our questions are based on what students say, with no agenda for where we think or hope their answers might lead. *How* our questions follow students' thinking, though, depends both on what the student says and on our goals—which can shift in an instant. Sometimes we might ask a general question to understand what a student has just said; other times, we might focus on something particular in her explanation that we want to clarify or that we want others to notice. We also might decide, on the spur of the moment, that students would benefit most from talking with one another about an idea that has come up. There are a variety of paths to take during a student's explanation, and no question is necessarily "the right question." It depends. We guess at it. We improvise. And we get better with practice. That's why some researchers describe the kind of discourse that follows students' thinking as "ambitious" teaching (Kazemi, Franke, and Lampert 2009).

## Starting the Number Talk

Number Talks begin with an invitation to share. Some variation of "Is anybody willing to raise their hand and say what they think the answer is?" opens the door to a variety of answers. Once answers are gathered, the teacher then asks if someone is willing to explain how they figured out their answer and why their strategy makes sense. Even though these initial questions may be easier to ask, they are important for helping us get our bearings as listeners and helping to position students as the "authors" of their strategies. As Johnston (2004) points out, questions like "How did you figure that out?" invite "a sense of agency as part of the [student's mathematical] identity" (31).

## Pressing for Conceptual Explanations

Once a student has started explaining his strategy, a central and crucial role for teachers is to "press" for why his strategy makes sense. It's not enough for students to describe what they did, even if it is a viable strategy and even if their strategy makes perfect sense (to us). They need, always, to explain *why*. In these "conceptual" explanations, students bring meaning to the numbers and operations they have used. When students explain conceptually, they "own" the mathematical ideas they've used, and their understanding becomes deeper and more connected. And every time they explain, they get better at learning how to make their explanations clear, concise, and mathematically accurate.

To encourage conceptual explanations, our questions need to be specific enough to draw out the meaning in particular parts of students' strategies, so we press for justification, clarification, and elaboration. But we can't press on *every* part of a student's strategy because too many questions can break the flow of an explanation and bog down the Number Talk. Deciding what to press on—and what not to—is a matter of experience and judgment. We press more gently, for example, when Number Talks are first getting started (see "graduated pressing" in *Making Number Talks Matter* [Humphreys and Parker 2015, 18–19]) so that students can get used to the idea that our questions don't mean something is wrong. As students become more confident in explaining their reasoning, we press differently. This all means that teachers are thinking continuously during a Number Talk, alert to what to press on and what to leave alone.

Here, we offer examples of teachers' decisions to press on particular aspects of students' explanations during Number Talks.

In our first example, Jay's Algebra 2 class has just mentally calculated 7.4 – 2.58. As a part of his strategy, Paul invokes a rule about significant digits that he learned in chemistry to justify truncating 2.58 to 2.5, which changed the problem to (the much easier) 7.4 – 2.5. We join the Number Talk as Paul explains that he "removes" the decimal points, subtracts 25 from 74, and gets 49. Instead of just accepting the answer and moving to the next step, though, Jay presses Paul to explain how he figured that out.

Jay presses Paul to explain how he knew that 74 – 25 is 49.
http://sten.pub/dd21

Even what might appear to be such a small question communicates that *how* Paul got the answer—not the answer itself—is what matters most to Jay. We also notice that Paul, without prompting, explains why his method makes sense. We've seen this during other Number Talks; once students expect to be asked to explain, they automatically follow a description of their method with, "... because ..." And "explaining why" gradually becomes the norm.

Our next example comes from a fourth-grade Number Talk. Danni has calculated 62 – 29 and begins by telling Hailey that she "decomposed the subtrahend." Hailey listens without comment, recording as Danni explains that she split the 29 into 20 and 9, and then split the 9 into 7 and 2. At that point, Hailey presses for elaboration: "Can you tell everyone why you chose to split up your number this way?" This question positions Danni as having made a strategic decision for how to decompose 29 and gives her the chance to put her reasoning into words.

Hailey "presses" on an important part of Danni's strategy.
http://sten.pub/dd22

It's also worth considering what Hailey might have asked but didn't—how, for example, Danni figured out 60 – 7 or even 53 – 20. Those questions could have been interesting, as Jay showed us, but they also could have made Danni's explanation longer, and there were a lot of kids waiting to share. Instead, Hailey focused the class's attention on one important part of Danni's strategy: *how* Danni decomposed the subtrahend, and *why* she did it in the particular way that she did.

In another video excerpt, we join a third-grade class that has just calculated 81 – 26. Isa has used a strategy she calls "constant difference" to transform 81 – 26 into 85 – 30 by adding 4 to both the minuend and subtrahend. One important element in Isa's thinking is how she chose to make use of this strategy—and why. So Hailey focuses her question on the 4.

Hailey focuses on why Isa added 4 to both numbers.
http://sten.pub/dd23

Notice also that Hailey didn't say, "Can you tell me?" and she didn't even say, "Why did you add 4?" Her use of the word *everyone* reminded Isa and the class that her explanation was important for everyone.

In our final example, Nisha's eighth-grade algebra class has just calculated 32 × 12. Logan is the first to share his strategy, explaining that he divided 12 by 2, and then added 32 six times to get 32 × 6. Nisha listens and records Logan's thinking without comment for a while. Because his method involves many calculations, Nisha doesn't dare press on every little thing, so she has to decide. She doesn't ask

Logan, for example, how he knew that 32 plus 32 is 64, or about how he added the two 64s. She asks where the third 64 came from (which was important to the strategy) but not how he calculated 128 plus 64. She does ask how he calculated 192 × 2, but not why. Then Nisha stops one more time when Logan says, "90 + 90, *or* 90 × 2." Asking, "Well, which one? Is it 90 + 90 or . . . ?" sends a message: "It matters what *you* did."

Nisha presses Logan for elaboration and justification.
http://sten.pub/dd24

In hindsight, *why* Logan multiplied 192 by 2 was probably more central to his strategy than *how* he multiplied. But in the middle of a Number Talk, we don't have the luxury of hindsight. It's hard to make instantaneous judgments about what is most important to focus on in a student's explanation, and it's impossible to always make the "right" decision (if there is such a thing). But if we have time to reflect on our Number Talks, we may realize that there was a better question we might have asked, and we can tuck away that question for another time.

Deciding when to press on students' thinking—and why—takes lots of practice and reflection. Learning to press on specific parts of an explanation without guiding students' ideas takes persistence and practice. Even when we ask mostly authentic questions, though, slipping back into known-answer questioning patterns can send mixed messages.

## Mixed Messages

Authentic questions send inherently different messages than known-answer questions. While authentic questions encourage students' expression of their own mathematical ideas, known-answer questions generally seek responses that originate outside students' personal sense making. So, during Number Talks, even when we ask mostly open, authentic questions, the known-answer questions that slip out involuntarily can undermine students' budding ownership of mathematical ideas. To help you be on the lookout for mixed messages, we discuss three kinds of teaching moves that inadvertently keep students dependent on us as their source for mathematical ideas: funneling sequences, leading questions, and filling.

### Funneling Sequences

A funneling sequence of questions follows a familiar pattern that is easy to slip into. The teacher asks a known-answer question; then, if the answer is wrong, the teacher embarks on an agonizing sequence of questions to get the student to say the right answer. This questioning pattern is called "funneling" because the questions converge on an answer the teacher seeks. Here's a short example from Wood (1998, 171):

The teacher has just asked Jim to give the answer to 9 + 7.

**Jim:** 14.

**Teacher:** OK. 7 plus 7 equals 14. 8 plus 7 is just adding one more to 14, which makes _____? *(voice slightly rising)*

**Jim:** 15.

**Teacher:** And 9 is one more than 8. So 15 plus one more is _____?

**Jim:** 16.

Through this series of if-then statements followed by fill-in-the-blank questions, Jim eventually says the right answer. The pattern the teacher used to "help" Jim was the teacher's logic, the teacher's thinking, and the teacher's idea. Not Jim's.

## Leading Questions

Leading questions restrict students' expression of their own mathematical ideas. Some questions that require a "yes" or "no" answer qualify as leading questions, as do "fill-in-the-blank" questions (like the teacher asked Jim) and "this-or-that" questions. Here's an example of a this-or-that question: "Were you adding extra eighteens or were you adding extra twelves?" Leading questions depend on and follow the teacher's—rather than the student's—line of reasoning. One way to know whether we have asked a leading question is by how students answer. If they give single-word or short-phrased answers, we have probably just asked a leading question.

## "Filling"

In "filling," we get ahead of or take over a student's ideas. Assuming we know what a student is thinking, we finish their explanation, make a connection, or expand on an idea that the student hasn't yet expressed. Filling is different from fill-in-the-blank questions where students are expected to give an anticipated response to a well-choreographed statement posed as a question. With "filling," the teacher finishes or embellishes the student's ideas. Here's an example from a Number Talk in which eighth-grade students had calculated 15 × 8. Joe is defending his answer of 120.

**Joe:** I cut the 8 in half to get 4. And 4 × 15 is 60.

**Teacher:** So you cut the 8 into two parts and you thought about 15 × 4 is four of these 8s and you had another 15 × 4?

**Joe:** Yeah.

Joe was only half done. But instead of pressing Joe to articulate the rest of his strategy, the teacher stumbled into a cultural trap: assuming how Joe had finished his strategy, she jumped ahead and articulated (what she thought was) *his* idea in *her* words. It doesn't matter that Joe said, "Yeah." What else would he say?

There was much lost in this short but significant interaction. Joe never got a chance to fully explain his thinking. But the teacher lost too. She lost the opportunity to learn how Joe was thinking; she lost the chance to help Joe get better at communicating clearly. Most important, though, she lost the chance to send the message to Joe and the rest of the class that it really was his thinking that she was interested in. It's ridiculously hard—for all of us—to let go of the familiar questions that lead, guide, and funnel students' responses. Teachers are supposed to help, but for too long, helping in math has meant doing the thinking for our students. When we ask an authentic question, however, we open ourselves to how students are thinking. This means that we have to listen carefully to what they say.

Melissa, a preservice teacher who had just completed ten Number Talks with her high school students, talks about how important Number Talks have been in helping her learn to listen:

Listening to students.
http://sten.pub/dd25

Melissa suggests that, as a new teacher, she hasn't yet gotten the habit of departing from a lesson plan to follow a student's idea down a path where the lesson might need to go. But she's thinking about it.

## Helping Students Learn to Engage with Each Other's Ideas

The questioning practices we've discussed in this chapter support students as they learn to make sense of quantities, properties, and numerical relationships. When students respond to authentic questions, they develop ownership of mathematical ideas and can begin to feel personal power as sense makers and mathematical thinkers. This, alone, is a worthy purpose of Number Talks. Lately, though, we've become more and more aware of something else Number Talks can offer beyond numerical fluency and flexibility. They can help students learn to actively engage with each other's ideas.

When we wrote *Making Number Talks Matter* (Humphreys and Parker 2015), we said, "Something wonderful happens when students learn they can make sense of mathematics in their own ways, make mathematically convincing arguments, *and build on the ideas of their peers*" (5, emphasis added). At the time, we believed that students who are comfortable with justifying their thinking and hearing others' justifications would naturally begin to ask questions or make comments that build on those responses. This can—and does—happen in some classrooms. But in many other classrooms, students sit silently during a teacher-student interaction, even when the exchange may be full of interesting (or confusing) ideas. We've come to understand that many students are unlikely to become actively involved with each other's ideas during Number Talks unless we explicitly invite and encourage them to become part of the discourse.

It's not that we haven't tried to get students to engage with each other! In *Making Number Talks Matter*, for example, we suggest that asking, "Does anyone have a question for ____?" after a student shares a strategy can rouse student interest (2015, 12). But as we've watched and enacted many Number Talks since then, we've noticed that this question can have the same effect as the familiar "Any questions?" at the end of a lecture: dead silence. And when students do respond, their questions are more like comments in disguise; for example, "Why did you do ____ instead of _____?" These experiences have led us on a quest to learn how to coax students into a new role as math students: interacting directly with their peers in what is called "dialogic" discourse.

In what follows, we offer a glimpse into what you might see when students begin to *own* the mathematical ideas under consideration and feel a bit of control over where the discourse might lead. These student behaviors have a look and feel that might be unfamiliar and even uncomfortable at first, but we view them as emerging student agency. Here's what your class might look like as you begin to share with your students the ownership of mathematical ideas and the direction of classroom discourse:

**As students begin to see themselves as active agents in the discussion of mathematical ideas, they will begin to**

- make comments or ask questions, unsolicited by you—that is, *not* after you have just said, "Any questions or comments for ___?" Wait time helps encourage this.
- speak directly to another student in the class, unprompted by you, and maybe get into back-and-forth discussions about the ideas.
- finish your sentences or even interrupt what you are saying to supply ideas of their own. (See, for example, Chapter 1, "A High School Number Talk: The Case of Frosty," where Frosty interrupts Cathy.)

**If you've been working on mistakes and wrong answers (see Chapter 2, "Mistakes") and your class culture around mistakes and confusion is shifting, students may start to**

- share ideas they aren't sure about.
- share how they got started with a strategy but can't figure out where to go from there.

- say, "I'm confused."
- share an answer they know is wrong and either explain or want to find out where they went wrong.

**Here are some teaching moves that can help generate students' engagement with others' ideas:**

- First, talk to your students about the changes in Number Talk conversations you are trying to make. Explain what you are trying to do and why. Tell them you are learning, too, and that you will make mistakes. Express a genuine desire for them to become interested in each other's ideas. Students may be used to whole-class discussions in other classes but often not in math.
- Early on, point out behaviors that support interactive discourse when they happen, or soon afterwards. This helps students develop an image of what that behavior looks like and why it is important. For example, in Chapter 1, Cathy talks to the class about how important it is that Frosty shared his confusion about the answers being offered.
- Move (physically) to the side of the classroom when students start to engage with each other. Because you're recording during Number Talks, you can feel tethered to the front of the room, but once a conversation starts, you can unobtrusively move to the side.
- Encourage students to engage with others' ideas by asking, "Who thinks they can put Michelle's idea into their own words?" Some students, though, shy away from having their ideas in the limelight, so we try to err on the side of caution by first asking something like, "Michelle, I want to be sure everybody understands what you said, so is it okay if I ask everyone to think about it?"
- Redirect questions about a strategy to the student who owns the strategy. Students are in the habit of directing their questions and comments to the teacher, even when they relate to someone else's idea. In that situation, try saying something like, "Why don't you ask Jorge about that?" Redirecting the question to Jorge supports students in learning to speak directly to each other.
- Reflect a student's question to the class without answering it. Example: When a student asks a question like, "Is that because the two numbers are odd?" toss the question back to the class by asking, "What do you think about that?" And when an idea is worth digging into a bit, ask students to put their heads together to discuss the question.
- There are times, during Number Talks, when you might want to ask students to work together to examine an idea. And again, some students shy away from having their ideas in the limelight, so let your knowledge of your students and your professional judgment and intuition guide you.

## General Tips and Suggestions

Here we offer a few general suggestions to help you make your peace with asking questions during Number Talks.

### Expect to feel uncomfortable!

We don't always know how to ask authentic questions—and how could we, with so little experience? As Thompson and her colleagues (1994) point out the following:

> When we move our instruction to deep conceptualizations of situations, we also move away from the domains of discourse with which we feel most comfortable—established methods for deriving numerical solutions. Instead, we move toward domains of discourse that emphasize "how you think about it"—domains few of us have explored and too few students have experienced. (89–90)

Your students are likely to feel uncomfortable that you are asking them how they think rather than telling them what to do. So there's not much we can do other than talk with them and move forward through the discomfort—learning together. We're rarely sure in the moment that we've asked the right question or pressed at the appropriate time, and this can keep us feeling a bit off balance. It takes time to become more comfortable with uncertainty. And it takes time to begin to see the results of the shifts we're trying to make in our questioning.

### Ask. Even when you think you already know what a student means, ask.

Many of us have learned to ask questions primarily when something is wrong with an answer or explanation, so asking "Why?" during a Number Talk, when the student's strategy is correct, complete, and clearly explained, can seem counterintuitive. It helps to remember that when we press students for explanations, their responses help everyone in the class understand. Also, although we may *think* we know why a student has used a particular strategy or chosen certain numbers, we don't *really* know until we ask.

### Things you can do when you don't understand what students are saying

It helps to pause and listen to the rest of the strategy so you know what's coming before you continue to record. If that doesn't help, then try asking a question like, "Can you take us through that again?" Here's an example where that particular question might have been useful. Aurora, an eighth grader, had calculated 8.4 – 3.75 by breaking the numbers apart, and Ruth was struggling to understand and capture what Aurora was doing:

Ruth is confused about Aurora's explanation.
http://sten.pub/dd26

Ruth, in reflecting on her decisions, says, "I wish I had asked Aurora to take us through her explanation again. That would have given her time to put her strategy more succinctly into words and it would have given me a chance to hear it again so I knew what questions I wanted to ask."

If, at the end of an explanation, you are still stumped, try saying something like, "I need some more time to think about your strategy. Is it okay if I think about it and get back to you?" Saying this conveys that their thinking is important to you and worth taking the time to figure out. Then talk with the student one-on-one at another time.

### When a student gets lost or stuck while explaining

When a student gets lost in an explanation, the first thing to do is *wait*, as we see Ruth do in this video when Logan, an eighth grader, forgets what he did while solving the problem 8.4 – 3.75.

Logan forgets what he did.
http://sten.pub/dd27

Logan first adds .35 to 8.4 to get 8.75 and Ruth presses to find out where the .35 came from. (Logan wanted the decimals in the two numbers to be the same.) He subtracts 3.75 from 8.75 to get 500 [sic], but then can't remember what he did next. After waiting, Ruth tries repeating what Logan did, thinking that this might help him remember what he did next. It doesn't. And so when Logan, looking a bit uncomfortable, again says he can't remember, Ruth asks if he wants more time to figure it out on his own or if he wants the class to think about it with him. When he says he wants more time, Ruth responds as if that's perfectly normal.

When an idea is complicated, or when students aren't used to explaining their thinking, they can get lost in their own strategies. So when a student gets stuck, offer him more time to think. Often, students will take that time and then raise their hand, having figured out what they did; other times, as with Logan, they won't. It's also good to acknowledge to the whole class that it can be hard to recapture an idea that's lost when everyone is focused on you.

**Don't forget to wait!**

Taking a deep breath and waiting gives you—and your students—a chance to think about what someone has said before you respond. Strategically, you can ask the student who is speaking to slow down a bit so that you can record her thinking, and then maybe stand back, look at the board, and think. This gives your students time to think, too. (See also Chapter 3, "Revising the Number Talk Routine: Rethinking Wait Time.")

Learning to avoid the closed, leading questions so familiar to all of us takes constant vigilance and practice. Our efforts, though, do not go unrewarded. Here, Tara talks about what she has learned about asking questions after doing just ten Number Talks with her high school geometry students:

Tara talks about learning to ask open-ended questions.
http://sten.pub/dd28

Asking authentic questions makes teaching unpredictable, yes, but also much more interesting. It's fascinating, fun, and sometimes downright baffling to find out how our students actually think. And how can we teach if we don't know what students are thinking?

CHAPTER

# 5 Nudging

## Helping Students Have Their Own Mathematical Ideas

Many of us learned that mathematics consists largely of rules and procedures to be memorized and practiced. Number Talks, however, are different. They open up mathematics to discovery and help students learn different ways of interacting with mathematics and with one another. They can help students come to care about the thinking of others and learn not to be content until they understand the mathematical ideas at play. Number Talks have the power to chip away at the prevailing culture of mathematics that students have come to know and rely on.

It can be really challenging, though, for students to let go of the algorithms that have dominated their mathematics instruction. It is hard for them to break free of the belief that there is one way to solve arithmetic problems when that's what they've learned over the years. Students have come to rely heavily on their teachers and textbooks for methods to solve problems, so it's no wonder that many of them are ill-equipped and perhaps even unwilling to step out on their own. We shouldn't be surprised, then, to be met with blank stares when we ask students to find "easier" ways to solve a problem or to find a strategy that "makes sense." And as middle and high school teachers are well aware, this problem only gets worse the longer students are in school. In this video, Tara, who teaches high school geometry, talks about her students just wanting to know "the rules."

Tara talks about how hard it is for students to get beyond standard algorithms.
http://sten.pub/dd29

In *Making Number Talks Matter* (Humphreys and Parker 2015), we express our belief that "As the numbers get larger in Number Talks, students who continue to cling to these algorithms will gradually realize for themselves that other methods can be much easier and more efficient" (165). But this does not always happen. We also recommend that teachers consider suggesting strategies they've "seen in another class" (165), which also can be helpful but, again, not always. And, when faced with a classroom full of students who suggest only one or two strategies during a Number Talk, it can be tempting to return to what's familiar and comfortable: demonstrating new strategies and then having students practice. Here, we discuss the trouble with this approach.

## Problems with Teaching Strategies Directly

Why not just teach strategies directly? Peter Johnston, celebrated literacy educator and author, observes, "Teaching strategies results in students knowing strategies, but not necessarily in their acting strategically, or having a sense of agency" (2004, 31). This is just as true in mathematics teaching as in literacy. Number Talks seek to help students learn to rely on their own ideas and mathematical understandings in order to solve problems flexibly and strategically. If, for some reason, their sense of agency in choosing and using their own strategies has been quashed, we need to find ways to rekindle their enthusiasm and confidence without doing the thinking for them. All students can think, but in math, they've too often learned, instead, to remember. The following video excerpts from a fourth-grade Number Talk help to illustrate what it looks like when students use ideas they don't understand.

We visited the class in mid-October. Hailey, the school's math coach, was working with the class to support their classroom teacher, who was new to Number Talks. Prior to our visit, Hailey had done just a few Number Talks with these students. For this Number Talk she had chosen 62 – 29, which we all thought would be well within the grasp of fourth graders who have made sense of subtraction, and we anticipated that most kids would subtract 30 and then put 1 back. But as the Number Talk unfolded, we were baffled, unsure of what we were seeing.

When Hailey first posed the problem, we were surprised by the number of different answers, noticing in particular that one of them (74) didn't make sense.

Hailey gathers answers for 62 – 29.
http://sten.pub/dd30

Then Colson, the first student to share, said, "The strategy I did was the one where you add 1 to both numbers." But changing 62 − 29 to the much easier 63 − 30 did not help him act strategically; instead, he slipped into using the subtraction algorithm he had been taught.

Colson uses the standard algorithm for 63 − 30.
http://sten.pub/dd31

After the Number Talk, Hailey met with Emily, Cathy, and Ruth to debrief the lesson. In this video clip, they discuss the number of answers and wonder about one strategy that was used by several students: adding the same number to both the subtrahend and minuend.

Emily and Hailey wonder about the number of answers and the "add one" strategy.
http://sten.pub/dd32

Then, as the Number Talk continued, we noticed that many of the strategies were unnecessarily cumbersome. One of the most jarring examples came late in the Number Talk, when Emily seems to have taken to heart a strategy called "Add Instead" and simply added instead of subtracting, as we see in this video.

Emily, literally, "adds instead."
http://sten.pub/dd33

And Emily wasn't the only one who did this. In the lesson debrief, we pondered what to do.

Hailey, Ruth, Cathy, and Emily discuss what happens and what to do when students are following procedures they've been taught.
http://sten.pub/dd34

While these examples are from only one Number Talk in just one class, they nevertheless allow us to consider the broader implications. The kinds of student responses we see here are not uncommon when students try to replicate methods they have been shown. Even when strategies are correctly applied, they can be cumbersome for solving problems mentally, as we see in this eighth-grade algebra class. First Grace divides a square into four parts for the four partial products. Then Hailey uses the distributive property.

Grace uses a commonly taught alternative method for multiplying 32 x 12.
http://sten.pub/dd35

Hailey also uses the distributive property.
http://sten.pub/dd36

Although both students' strategies result in the correct answer, they are cumbersome for mental computation. Deja, on the other hand, solves the problem easily by doing 32 × 10 and 32 × 2.

Deja makes it easy to solve 32 x 12.
http://sten.pub/dd37

When students figure out strategies that make a problem easier, as Deja did, they are empowered as "authors" of their own mathematical ideas. This is what we want for all students.

Watching Number Talks like these has led us to think more about how to help students generate their own ideas and develop numerical flexibility. And, since flexibility requires understanding how numbers and operations work, we purposely start with very small numbers that students can easily conceptualize. As they learn to strategically break small numbers apart or change the numbers to make a problem easier, they come to better understand the structure of numbers and operations. These understandings form the foundation for larger numbers.

We've come up with three approaches that just might eliminate some of the thorniest problems encountered during Number Talks—at all grade levels. If you choose one operation (such as multiplication) and start with the first approach before moving to the second—staying with the same operation—you might not need the third approach at all!

## Nudging for Flexible and Strategic Thinking

**FIRST APPROACH: "What could you do if you didn't already know the answer?"**
Use this approach when moving from Dot Talks to Number Talks or whenever you begin working with a new operation. Choose addition "facts" to 9 + 9 (e.g., 6 + 7 or, with subtraction, 17 – 9) or multiplication facts to 9 x 9.

With this approach, set up a regular Number Talk, but tell students that you're going to put up a problem they probably already know the answer to.

- Explain that the goal of this Number Talk is to figure out what they might do if they didn't already know the answer. Say, "Think about what you might try if you didn't know the answer. How could you use something you already know to make this an easy problem to do?"
- Then do the Number Talk as usual, asking questions like, "Who made it easy in a different way?"
- Record just like you do with any Number Talk.

An added benefit of this first approach to nudging is that students who don't know their multiplication or addition combinations now have a sense-making way to figure them out.

**SECOND APPROACH: "What could you do if you hadn't been shown how to do this?"**
Use this approach when your students are ready for slightly larger problems that can be solved using a paper-and-pencil algorithm. The goal here is to take standard algorithms "off the table" during Number Talks. Choose problems that you think are small enough to be within conceptual reach for your students to solve mentally. Don't be tempted to get too big too fast; your students will make more progress if you give them the chance to build their foundations slowly. For example, if your students have a lot of strategies for problems like 7 x 8, then try a one-digit by two-digit multiplication problem such as 5 x 18.

In *Making Number Talks Matter* (Humphreys and Parker 2015, 165–166), we suggest things to do when a student shares the standard algorithm, but we've had a nagging feeling that there might be a better way to address algorithms before students even use them. We say, for example, that when the traditional algorithm is offered the first time, we would "explain briefly what an algorithm is. From then on . . . we write *traditional algorithm* on the board" (165; italics in original). We've now come to believe, though, that waiting until algorithms arise is too late. Because of the various ways that teachers respond when students first use a standard algorithm during Number Talks, it's very possible that students get the message that their method isn't adequate or doesn't count as a "real" strategy. This second nudging approach is designed to nip this problem in the bud. By taking standard algorithms off the table right from the beginning, students are encouraged to think strategically about ways to make a problem easier to solve.

We usually say something like, "Here's a problem we've all seen before: 43 – 18. Most of us learned there's one way to solve this problem. But you know what? That's not true! There are lots of ways to solve problems like this, and sometimes they're way easier than the methods we've been taught. If you went to school in the United States, you probably all learned to solve the problem the same way. Would anyone want to tell us how to start this problem?" Continue this way for a few steps, asking each time for student input. Don't bother to explain why the algorithm works, because right now you are trying to nudge students away from their dependence on standard algorithms.

Then write a similar problem on the board and say, "Think about how you could make this an easier problem to solve mentally." Wait a few seconds and say, "Put your thumbs up when you have a way to make the problem easier." Wait until a lot of thumbs come up—and then wait some more. Record the strategies as they come up.

If a variety of strategies is still not forthcoming, then ask the class to engage together by saying, "Let's think about this together. How could we make this problem easier?" Continue to ask, "How else?" and "How else?" And don't forget to wait. You might also ask students to talk briefly together to come up with other ways to make the problem easy to think about.

### THIRD APPROACH: *Good First Moves*

This last approach, not really a Number Talk, is a whole-class activity designed to help students develop computational flexibility. The "first move" in a Number Talk is the very first choice students make about what to do with the numbers to help them get started on the problem (e.g., for 63 – 28, a student might decide to start with 63 – 30). This activity has three parts: generating ideas, working in pairs or small groups, and sharing out. It usually takes at least thirty minutes, and we have found it to be well worth the time.

Use this approach if students seem stuck on the same few strategies, if the same handful of students are always sharing, or if the standard algorithm is creeping back in.

Choose problems that lend themselves to a variety of strategies: for example, 73 – 28, 86 – 17, 38 + 63, 97 + 44, 12 × 16, 26 × 12.

**Part 1: Generating ideas**

- Say, "Take a look at this problem and think about how you might start to solve it. No need to figure out the answer! Just think about what you might do first, and why." As in any Number Talk, wait until most thumbs are up.

- Then ask, "Is anyone willing to share how you chose to start?" Asking students how they chose to start helps them understand that they are *making choices* during Number Talks. Let students know that you want them to learn to make strategic choices about where they start and how they proceed through any given problem.

- While collecting their suggestions, let students know you want to hear only their first step; then record. For example, in the problem 73 – 28, one student starts with 73 – 30, while another suggests 73 – 20. A third student says that he added 2 to both numbers, and yet another student says, "I broke the 28 into 23 and 5." The teacher records:

  73 – 30      73 – 20      73 – 28      73 – 28
                               +2   +2          23^5

- Continue to ask for and record potential first steps until you have three or four suggestions.

**Part 2: Working in pairs or small groups**

- Ask students to start with each suggested first move—they should use paper and pencil to keep track of their work—and finish solving the problem.

- Suggest that each time they carry out a step, they should look at the results and talk together about what might be a "good next move," and then continue until the problem has been solved.

**Part 3: Sharing out**

- When students have had time to solve the problem using each of the "good first moves," choose one of the suggested moves and ask if anyone is willing to share how he or she finished the problem. Record as in a Number Talk.

- Invite others who used the same first move but solved the problem differently to share what they did and why. Again, record.

- Continue with the other first moves as time allows. This process highlights the notion that strategic thinking at each stage of a problem is important in mathematics.

"Good First Moves" also helps students realize that even if they start with a particular strategy, they might change to a different strategy as they work their way through the problem. For example, in $16 \times 12$, a student might first halve and double to get $8 \times 24$. Then instead of halving and doubling again, she might find it easier to round the 24 to 25 and adjust; this would give her $8 \times 25 = 200$ and then $200 - 8 = 192$.

Figuring out new ways to help students develop flexibility has always been a goal for us, and we think these three approaches to nudging go a long way towards achieving that goal. When Number Talk strategies come from students themselves rather than being taught directly, the class culture can more easily shift from listening and practicing to doing and understanding.

CHAPTER

# 6 Dot Talks

We used to think that Dot Talks were only for the primary grades, but Sally Keyes, our friend and colleague, helped us realize that Dot Talks are the foundation for Number Talks—no matter what grade level you teach (see *Making Number Talks Matter* 2015, 14–17). Dot Talks provide access and a safe space, even for students who might otherwise be reluctant to talk in class. Everyone has ways to see the dots on a card, and students fairly quickly take up the challenge of seeing them in multiple ways. Dot Talks can help build a sense of community in mathematics classrooms and allow even reluctant learners to become part of that community.

The teaching decisions are similar with Dot Talks and Number Talks. In this chapter, we use video excerpts from a fourth-grade Dot Talk to bring those decisions into sharper focus and consider some common questions that arise around Dot Talks.

Starting the Dot Talk.
http://sten.pub/dd38

## Common Questions About Dot Talks

### Q: *How do you choose which dot card to use?*

**A:** We choose dot cards that are within reach of students, with arrangements of dots (or other figures) that call out for a variety of ways to "see." A mix of cards with numbers ranging from six to ten or so is a good place to start with elementary children—and just as good a place to start with high schoolers.

We don't go into a Dot Talk with an agenda of what we hope students will see on any particular dot card, but we anticipate ways they might see. When we select dot cards, we choose arrangements where we're genuinely curious about what students will see, knowing we can't anticipate all of what students might see or do. This way, we can enter each Dot Talk with wonderings: How will students see the dots? Will all students engage? Who will want to share what? And when? What will we learn about students? What will they learn from each other? With Dot Talks, as with Number Talks, there's much for a teacher to be curious about! You'll find a selection of dot cards in Appendix B.

### Q: *Why do you leave the dot cards up the whole time? I'm used to showing them quickly.*

**A:** Dot Talks are similar to Quick Images. In Quick Images, though, the cards are shown quickly and then removed, while in Dot Talks the cards remain visible throughout. Both Dot Talks and Quick Images help students learn to look for small clumps of dots or images that they "know" without having to count one by one. According to Carpenter and his colleagues (2017),

> For small collections of objects, children can directly perceive the number of objects in the collection without having to count. This process, called *subitizing*, is thought to play a fundamental role in the development of children's basic concept of number. (16; italics in original)

Dot Talks give students an opportunity to subitize without the need to be quick. In this video clip we see Emily subitize when she sees five "like on a domino."

Emily is subitizing.
http://sten.pub/dd39

When dot cards remain visible, though, students can also learn to see the dot arrangements in more than one way and watch as others describe what they see.

## Q: *Why don't you use the equals sign with dot cards when recording what kids are saying—for example, "That equals . . ."?*

**A:** Unlike when we're doing Number Talks, we don't use the equals sign during Dot Talks for two main reasons. First, multiple equals signs can make it harder for students to see what others see. If we did use equals signs, then when Addie says in this video, "I did 3 plus 2 equals 5, then 5 plus 2 equals 7," we would record the following:

$$3 + 2 = 5$$
$$5 + 2 = 7$$

Since we don't use the equals sign, Ruth records just this:

$$3 + 2 + 2$$

Ruth records Addie's way of seeing.
http://sten.pub/dd40

The second reason not to use the equals sign in Dot Talks is to help all students realize that expressions themselves can be "answers." Writing an expression like 2 + 3 + 2 can head off a misconception that many students carry with them into algebra: that quantities are only represented by a single number (in this case, "7") or variable. In this Dot Talk, the ways students saw the seven dots were represented by different expressions: 3 + 2 + 2, 4 + 3, 5 + 2, 2 + 3 + 2, 3 + 4, and 3 + 3 + 1.

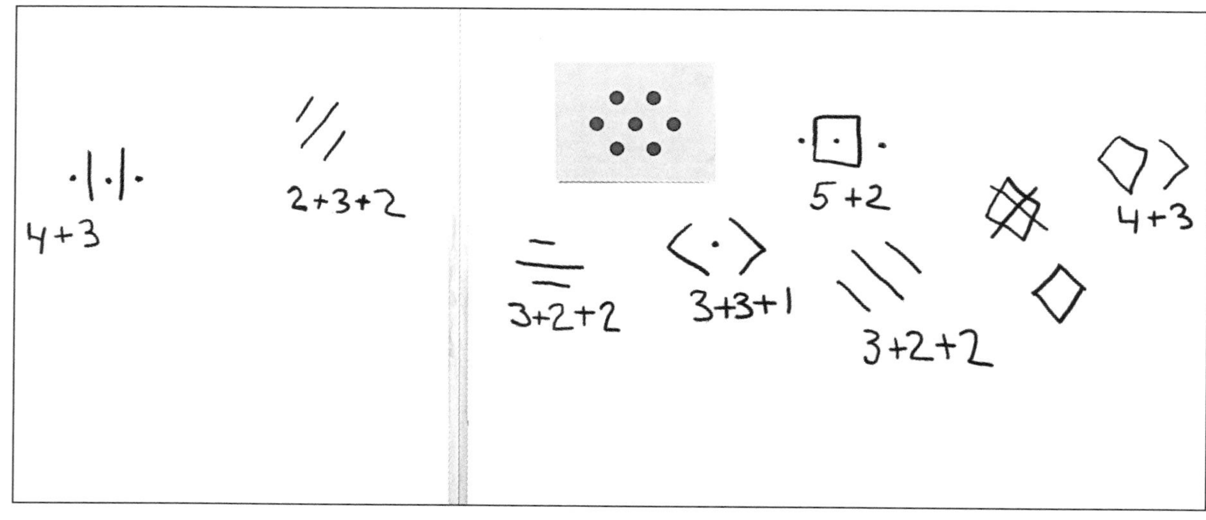


**FIGURE 6.1**
Screenshot of whiteboard in EL2.

## Q: *What do you do when students don't use correct mathematical language?*

**A:** When students don't use correct mathematical language to describe what they are seeing, the teacher's role is to use appropriate mathematical terms without making a big deal of it. Students will gravitate to the correct vocabulary when they understand it. In this video, when Elijah says he sees a diamond, Ruth records a rhombus as she says, "You saw a rhombus." Then in the following video, she calmly introduces the term *vertical* when Micah doesn't have the words to describe two dots, one above the other.

Elijah says "diamond" and Ruth records, saying, "rhombus."
http://sten.pub/dd41

Micah struggles to describe two dots and Ruth casually introduces the term *vertical*.
http://sten.pub/dd42

Dot Talks provide a rich opportunity for students to begin to use mathematical language to describe their ideas. The mathematical language comes not from a vocabulary lesson but from a natural need to communicate ideas more clearly and efficiently. Students want to know terms such as *diagonal, horizontal, top, bottom, left, right, hexagon, triangle, trapezoid,* and so forth as they learn that using the language of mathematics can make it easier to communicate. This helps English language learners, too, who can be part of conversations they want to engage in, with ideas that are accessible.

## Q: *Why do you record with lines instead of dots?*

**A:** Some teachers reproduce multiple copies of a dot card and circle groups of dots to indicate what each student sees. We use line segments as a more abstract way to represent what they see in order to help students visualize where the dots are *represented* in the line segments. This might not be appropriate for very young children, but by third grade most students can do this with ease.

Ruth records Addie and Mica's ways of seeing.
http://sten.pub/dd43

Ruth, Hailey, Emily, and Cathy discuss how we record during Dot Talks, and why.
http://sten.pub/dd44

## Q: *I've been using the same pen to record all of the ways students see the dots. Why do you change colors?*

**A:** We use a different color each time another student shares so that their story (or way of seeing) stays intact. When the pen color remains the same, the chart paper or whiteboard ends up covered with numerical expressions, and the individual ways of seeing can be lost. Color helps keep the strategies separate, making it easier, then, to look across strategies.

## Q: *Why don't you write students' names by their strategies?*

**A:** This is a question we've talked about a lot. Many teachers write students' names next to the strategies they have shared, and we have, too, in the past. Pairing strategies with names can help students see that people make sense of math in different ways—in their *own* ways—and this personalization can help the class develop a sense of camaraderie and community. Over time, though, we've become concerned about possible adverse effects of attributing strategies to particular students and are now rethinking this practice for both Dot Talks and Number Talks.

Our biggest concern is the potential effect of unwanted attention. While some students like having their name on the board during Number Talks, it's not true for everybody. In some Native American cultures, for example, it is not desirable to be in the spotlight. Also, students who are naturally shy may hesitate to share if it means that their name will be on the board. For these students, seeing their names next to a strategy can jeopardize a top priority of Dot Talks (and Number Talks): to provide a safe haven for *every* student to participate.

Another concern is how a student might feel about a mistake on the board next to her name. Helping students learn to see mistakes as natural and unavoidable is a top priority of Number Talks (see Chapter 2, "Mistakes"), but students carry with them a lot of emotional baggage from their prior experiences. Mistakes in math class are, therefore, fraught territory where we need to tread carefully. Horn (2017) points out, "Mathematics classrooms are particularly burdened with social risk" during the already socially risky time of adolescence (4). When a student gets the courage to share a strategy, only to realize that it is flawed or unnecessarily cumbersome, it can be mortifying to have their name up there for all to see.

Another issue, possibly more a wondering than a concern, is around ownership of a strategy. If Jessica is the first to express a particular strategy, it becomes "Jessica's strategy" because she was the first one to articulate it—regardless of how many other students did the same thing. We aren't sure how much of a problem this really is; some students mind, while others don't. But we still wonder.

While it's true that writing students' names next to strategies can make them easier to refer to, we have found that students continue to notice, reference, and build on each other's work, even when no names are attached. They also readily and, in the case of Dot Talks, eagerly share their ways of seeing. Besides, Dot Talks and Number Talks don't need students' names to create a sense of community. For us, the concerns about pairing names with strategies outweigh the benefits, so we now choose not to do so ourselves.

## Q: *Why don't you look at students while they're explaining how they see the dots?*

**A:** We're now wondering the same thing and think there's a better way to get at what we're after here: students who can communicate clearly, efficiently, and mathematically. Before we describe our "better way," though, let's first take a look at what was behind Ruth's choice to not look at Damien. This next clip shows the exchange between Damien and Ruth, and the following clip shows Hailey, Emily, Cathy, and Ruth talking about it.

Damien explains what he sees as Ruth listens without looking at him.
http://sten.pub/dd45

Ruth looks away to nudge students to use words for what they see.
http://sten.pub/dd46

Dot Talks help students learn to communicate their thinking in words. Yet almost always students show us what they're thinking with arm movements, using very few words. So, if we take our cues for what to record from their arm motions, will students then have a reason to avoid putting words to their ideas? Using arm motions can actually help some students think of words to say. For example, picture a child making vertical arcs with their arms while saying, "around the outside." Arm motions can also help other students notice where the dots are in the teacher's recordings. Looking away feels odd, though, so we think it would be better to let students use arm movements whenever they want to and push, gently at first, to encourage them to also put words to their ideas. Ask questions like these: "Can you think of a way to tell us where the dots are?" "What if we couldn't see you—what words might help?" "How can you describe in words where those dots are?" While this kind of probing is done initially with a light touch, we gradually begin to press harder until the culture finally shifts, students realize they're all in this together, and everyone asks questions until they understand.

## Q: *I can't stand watching my kids struggle to get their ideas out. Do you have any suggestions that might help?*

**A:** We try our best to listen carefully to students, trusting that they'll find ways to communicate their thinking. And we try *not* to fill in or finish their ideas or comments for them. Explaining what they see on a dot card is a challenge for many students, especially when they are just getting started. Even though Dot Talks are about what students see, learning to communicate with precision isn't easy. For most of us, it's a lifelong challenge. But Dot Talks go a long way toward helping students learn to communicate their ideas clearly.

Whether teachers are working with elementary students, high school students, or in higher education, Dot Talks are a safe place for students to begin to talk with each other. Explaining isn't easy for a lot of students. In this video Gloria struggles to explain how she sees the dots. But Ruth waits patiently, knowing that Gloria needs the practice and trusting that she will be able to explain what she sees.

Gloria explains what she sees.
http://sten.pub/dd47

It's tempting to fill in words for students who are struggling or finish their explanations for them. But when we do, we send an implicit message that we don't think they're capable of talking for themselves. Students get better at explaining over time when given ongoing opportunities to do so, and Dot Talks provide just such an opportunity. Ruth, in reflecting on the interaction with Gloria, says this:

> It was difficult to wait as Gloria searched for words. And as I watched her continue to struggle to tell me which diagonal she was seeing, it felt like a long, long wait. And I wondered: Was I losing the other kids? Was Gloria feeling stress? Not knowing, I attempted to take the heat off of Gloria and put it on myself by saying, "I have a question that might help me." But I wasn't at all sure I made the right call when I asked her which side the diagonal started on. I hoped, though, that asking about a right or left orientation might help Gloria describe where she saw the diagonal, and maybe even give other students an idea for how they might use positioning words to express what they see.

With Dot Talks, as with Number Talks, we often say things in the moment that we wish we could take back. But we're all learning something new here, and we're always trying to get better. Thankfully, students are forgiving, especially when we share with them what we're trying to do, and why.

## Q: *My students love Dot Talks, but I feel pressure to move on. How do I know when it's time to move on to Number Talks?*

**A:** Nearly all teachers feel this tension, which is even more pronounced at the secondary level. Staying with Dot Talks long enough to build a solid foundation for Number Talks is key. Many considerations go into deciding when to move on. Are Dot Talks still helping to engage everyone? Are students getting better at explaining what they see? Are they taking up mathematical language when appropriate? Are they seeming to revel in the fact that there are so many different ways to see the same thing? Are they getting the idea that they can see math in their own ways? Have they come to trust that they won't be called on until they're ready to share? Are they aware that they're part of a community where everyone counts? Are they feeling safe?

We want Dot Talks to be engaging and fun for students. And we want to bring this same level of engagement to Number Talks. It can help to step back and let students see what they're accomplishing through Dot Talks. You might want to engage the class in listing the mathematics they are using; both you and your students will probably be surprised at how many things they come up with. Students, for example often mention words they have learned for shape, orientation, and location. Terms such as *trapezoid*, *symmetry*, *vertical*, and *horizontal* come up frequently. Teachers can supply the ideas that students don't think of, such as visualizing, subitizing, and using equivalent expressions. It's also good to talk about what else is being learned; students sometimes say things like, "We don't all see things the same way, even in math," or "Communicating ideas is getting easier," or "People are talking more." This helps both teachers and students see Dot Talks as worth the time invested. It's also helpful to remember that the dispositions students develop during Dot Talks will lead to higher levels of participation as we move into the other areas of mathematics.

## Q: *I love what Dot Talks have done for my special needs students!*

**A:** This is not a question, but it's something we hear a lot. So many students find their mathematical voices through Dot Talks, and special needs students are no exception. Dot Talks can have an empowering effect on students who, for a variety of reasons, are often left out of the conversation when it comes to math.

There are things for everyone to learn from Dot Talks—how to communicate more clearly, how to use the language of mathematics, what it's like to listen carefully to another's idea, how to look for similarities and differences, how to push beyond your original ideas, how surprising it can be to see what others see, how to contribute to a productive learning community, and lots more. Dot

Talks are not just for elementary and support classes; they are good for all teachers and students—elementary through university.

## Q: *What might it look like to have students reflect on the different strategies shared (a question we've been asking ourselves)?*

**A:** We've been wondering what it might look like to use wait time when Dot Talks are coming to an end by asking students to look across strategies to think about what they notice or wonder, as we suggest doing with Number Talks (see Chapter 3, "Revising the Number Talk Routine: Rethinking Wait Time"). Helping students learn to look for similarities and differences between approaches is important, and we've been thinking more deeply about how we might use Dot Talks to begin to build a disposition for this.

Our first approach, always, would be to provide a space for students to look at the different ways of seeing and think about what they notice by asking a question like, "When you look at all of these ways of seeing, what do you notice? Or what do you wonder about?" Then we would allow time for them to look and indicate with their thumbs when they have something to share. We anticipate that students will, on their own, notice similarities.

But there might also be times when we want to focus their attention on something they haven't noticed. In this case we might, after first waiting to hear what students have to say on their own, ask them what they notice about what's the same or different between two or more specific strategies or ways of seeing. We use this sparingly, though, to keep the focus on students' own ideas.

Dot Talks lay the groundwork for helping students learn an essential lesson to bring to all of mathematics: no matter what the problem or situation is, not everyone sees it the same way. Dot Talks stimulate a natural sense of curiosity about the different ways others see the same picture. And this curiosity carries into Number Talks when students want to know how others have solved a problem that they, themselves, have been thinking about.

We hope this chapter will encourage teachers to slow down and spend quality time with Dot Talks before moving on to Number Talks and then return to them whenever they might be helpful. A deep dive into Dot Talks pretty much guarantees an enlightening learning experience for teachers and students alike.

After a recent Community Math Night where parents participated in Dot Talks and Number Talks, we asked participants to reflect on the most surprising thing they had learned. One parent wrote, "There's no reason to be afraid of math!" while another wrote, "Fear of math can be erased even after many years." Dot Talks are welcoming. They're a safe space where students, even those who might otherwise feel resistance when it comes to mathematics, can participate freely. And as such, they provide a way to begin to make mathematics more inclusive for all.

# CHAPTER 7 Safety

## Thoughts from Ruth

Ensuring that all students feel safe enough to want to talk is at the very heart of Number Talks. And, while all teachers want their students to feel safe, there are different ideas about what makes for a safe learning environment. In this chapter, I explain why one of our ways of making Number Talks safe is to *never* ask students to talk involuntarily.

From early in my teaching career, I've promised students on day one that I would try very hard to never put them on the spot by asking them to talk in front of others unless they were volunteering. As I explain here, this has a lot to do with my personal experiences as a learner. But, over the years, many students have thanked me for this. And teachers have told me that their students tell them that knowing they won't have to talk is what makes them want to talk.

My resolve to turn over to students the decision of whether and when to talk runs counter to some routines that teachers have come to believe in and practice and counter to what some teachers are being asked to do by their colleagues and administrators. But there are reasons why we made this decision, and reasons why it is important to the success of Number Talks. Sharing what brought me to this place will help me explain why.

## Confessions of a Learner

I'm shy. I've always been uncomfortable in situations where I might be asked to talk at any given moment. It's not that I don't have ideas to share. It's that whenever I've been put on the spot to talk, I've had a hard time even thinking. Having to talk when I don't feel ready puts me in a miserable state and pretty much ruins any learning experience. I'm fully aware of the irony of this. After all, since leaving the classroom, much of my work has involved giving talks throughout the country to large groups of people—parents, mathematicians, mathematics educators, administrators, and business and community leaders alike. I've become comfortable speaking in public, and nearly always really enjoy it, but only when I know ahead of time that I'll have to do it. For me, it's one thing to talk when I choose to and feel like I have something to say and another thing altogether when it's not my choice. I enjoy learning, and when I can decide freely whether or not to talk, I thrive in most learning settings. But when I'm not in control of the decision, I suffer.

I'm sometimes a really slow learner, too. My anxiety takes over when I'm asked to talk about something I'm expected to know . . . but don't. If I have a theory or a hunch, I don't mind sharing it. But ask me to engage in discourse about something important when I've had only a limited time to digest the ideas, and my feelings of shyness and inadequacy take over. And when I'm asked to talk for others, things only get worse. I so clearly remember being part of a small working group during a National Advisory Board meeting where we were given a task to work on and I was randomly assigned the role of "reporter" for the group. I felt sick to my stomach the whole time, knowing that I was going to have to represent my group when we reconvened. I remember frantically searching the entire time for what I was going to report, which made it nearly impossible to engage thoughtfully as a learner. I dreaded having to report out for the group, and when I did, I felt embarrassed that I had disappointed my colleagues (it didn't matter whether I had or not; I still felt that way).

I'm a grown adult who has been successful in my career, and if this practice is painful for me, I know there are students who must feel the same way. Without a doubt, we have an important job of teaching students what it takes to work successfully in small groups, but I know this can be done without ever putting students on the spot by asking them to talk when they might not want to.

In addition to being a slow learner, I'm a slow reader. I'm the only person I know who took a speed reading course in college and read more slowly at the end of the course than at the beginning! Having to digest something quickly and be ready and expected to talk about it is really stressful for me. I dread situations where, as part of a group, I'm given something to read and then expected to talk with others. Knowing others are waiting for me to finish reading usually results in my needing to read passages two and even three times to absorb their meaning. Then, often, I'm not ready to bring much to the table during ensuing discussions because when I feel like I have to rush to keep up, my anxiety takes over once again. Knowing I might be put on the spot can put me in such a state of agitation that I simply can't attend well to what I'm supposed to be learning or doing. From what I've read about brain research, it seems clear that my brain is downshifting during these times, in search of safety.

Dr. Robert Sylwester, a colleague and member of my doctoral dissertation committee, described the human brain as constantly taking in huge gulps of information. He explained the brain's capacity to sort to the foreground those things that are important to pay attention to and push to the background what's trivial. But Sylwester also explained how, under high levels of stress, the brain loses its capacity to sort the important from the trivial. He shared, as an example, a personal anecdote. He was on a cross-country flight when he first heard that President Kennedy had been assassinated. He remembered that a man sitting next to him on the plane was wearing a green tie. This was a trivial thing to remember forty-five years later, but it was permanently stored in his memory because it was taken in at a time of high stress when the brain loses its capacity to sort. Bob asked us to think about students who frequently experience high levels of stress in mathematics classrooms—stress that causes their brains to be unable to sort what's trivial from what's important to pay attention to.

Many experiences over the years have helped shape the teacher I am today. My career has been a decades-long process of shifting and refining my practice. Yet I continue to find that the one thing I am unable to compromise on is that promise I make to my students on day one. My resolve on this issue means that there are some currently popular practices I simply can't do because they violate that promise—and because they're just plain stressful for me and for learners like me. Here are just a few of those practices:

1. **Popsicle sticks**

   Popsicle sticks are sometimes called equity sticks, but I believe they are inherently inequitable because not all students have the same level of comfort with being called on randomly. I know that pulling Popsicle sticks (or another random name generator) is intended to make sure that all students are engaged. But I'm not at all convinced that this practice results in higher levels of engagement; quite the contrary. With Popsicle sticks, we might be able to make students participate (even if it's only by saying, "I pass"), but we can't make them engage. Engagement comes from being curious and wanting to figure things out. It comes from wanting to be part of a conversation—whether or not you talk. And, for me, whether using Popsicle sticks works or not, the cost is far too high. There are other ways to authentically engage students with mathematics. The use of Popsicle sticks puts real learning out of reach for some of us. When we're in situations where such practices are at play, we almost immediately begin a desperate search for the facts or ideas we might be called on to explain. Some kids even choose to disengage totally as a way to avoid being embarrassed in front of their peers.

Many teachers, though, tell me that they use Popsicle sticks and it's okay because they first create a safe learning environment where kids know that they can always say, "I pass." I don't doubt this practice is fine with some students. But for others of us, having to say, "I pass" is just another time to feel embarrassed in front of our peers. Just ask kids how they feel about Popsicle sticks, and they will tell you. As teachers, we're responsible for ensuring that students are engaged and have equal opportunities to be part of classroom discussions. Talking is vital to learning. But there are ways to encourage participation and keep track of who's talking without putting students on the spot when they wouldn't choose to be there.

**2. "Turn and Talk" where everyone is expected to talk**

I've frequently been in settings where we were asked to turn and talk with someone next to us, taking care to share the talking time. Those times, too, often filled me with dread. Every time, I would quickly ask the other person to go first, and while they were talking, I couldn't really listen because I was too distracted, searching for what I would say when it was my turn to talk. Afterwards, I would usually feel a flush of embarrassment, convinced that I had probably just shown my talking partner how little of value I had to say.

It's vital for students to talk with each other a lot while they're learning—and vital for them to feel safe doing so. My choice has been to never partner students in situations where everyone is expected to talk. Instead, I frequently ask them to talk with people around them and take care to ensure that everyone who wants to talk gets a chance to. Students rise to the occasion when they are asked to interact in this way. And it's a chance for them to learn healthy and respectful ways of interacting with others.

**3. Hand signals for "me too"**

Some students have learned to use the American Sign Language phrase for "me too" to indicate that they got the same answer or used the same strategy for solving a problem.

I've been asked to do this only a couple of times as an adult learner, and once again, I wasn't comfortable. Here's what's going on for some of us in the midst of showing hand signals. We're asking ourselves, *If I give the "me too" signal for a wrong answer, is everyone going to notice that I don't get it? Even if I'm confident, are there others out there who are feeling put on the spot? Who can I watch so I can piggyback onto their response*

> *and just do the same thing, since I trust they do get it? What if I disagree with nearly everybody—do I dare suggest a different answer? Am I being assessed right now?* Forget about thinking; our brains are in search of safety!
>
> When it comes to Number Talks, I've seen students who were about to share a different answer or strategy quickly put their hands down when they noticed all the hand signals happening around them for a previous answer or strategy. And there are always those students who feel pressure to just do what their friends or the "high status" students do. I wonder if the reticent student who has just gotten the nerve to share an idea, upon seeing no hand signals to agree with him, comes to doubt himself once again. One teacher has described hand signals as the equivalent of "likes" on Facebook.
>
> Even if hand signals seem benign, I don't believe they belong in Number Talks. A sense of voting, or showing popular ideas, can quickly undermine the kind of environment Kathy Richardson and I envisioned when we developed Number Talks. I'm guessing this practice was employed to help teachers (or their administrators) make sure everyone is participating, or maybe to assess where students are in the moment. Hand signals aren't a viable assessment tool, though, because it's too compelling for some students to just do what they see others doing. Hand signals can imply understanding where there is none. And even without the use of hand signals, teachers learn a great deal more about where students are in their thinking and understanding during Number Talks.

Allowing students to decide when to talk might seem risky, but I've found it to be a very powerful way of ensuring that all kids' voices are heard. Ironically, never requiring students to talk publicly results in high levels of student engagement. Other teachers have told me the same. Here are some of their voices shared spontaneously during a Mathematics Education Collaborative (MEC) webinar on January 18, 2018:

> One of my special education students told me that she has never wanted to share more than when she wasn't forced to because she was in control and she could share when she had something to offer. She also said that she felt for the first time like thinking differently was valued and something to be proud of that could potentially help others learn.
>
> **Jamie Souhrada, eighth-grade math**

> As a teacher, I used to think that by cold calling on students I was keeping them on task. I thought that since students had to be ready to respond, they would stay focused on our math discussions. My thoughts have changed. As a teacher in an alternative high school, I teach students who have profound issues with trust—particularly with trusting adults to keep them safe. This year, I am trying really hard not to cold call. Instead, I'm trying to say, "If you're willing to share, I'd really like to hear your thinking." I'm not always successful, but I'm slowly making the change. Students are liking the change in my practice. When I forget, they remind me!
>
> **Molly Coulter, high school math**

> I often have students who don't participate during the Number Talks come up to me immediately after or stop me in the hallway to explain their thinking. I love it!
>
> **Laura Becket, math coach, K–5**

> Not calling on students was hard at first. I remember how long the wait time felt, and how painful it seemed. I tried to implement this practice all of the time, not just during Number Talks. However, once the culture shifted, it was amazing! I had more students participating in every lesson and responding to every question than I ever had when I called on people. The class became truly student centered. It's as if the kids took control of the learning when I quit calling on them. Tough to explain, but it was a visible difference.
>
> **Kyla Gellerson, secondary teacher, instructional coach**

When students feel safe enough to engage, things get interesting, enlightening, and challenging. Number Talks can lure students into the world of mathematics and build their confidence and their belief in themselves as sense makers. No matter when a student decides to want in, Number Talks are welcoming.

I encourage everyone to try putting students in charge of when they talk. Pay attention to what's gained and what's lost. Bring your colleagues on board and work together to figure out what's possible. I'm confident that when you take a deep dive into Number Talks and their guiding principles (Humphreys and Parker 2015, 26–31), you'll begin to see the many benefits and rewards that come with this special kind of spontaneous and authentic student engagement.

CHAPTER

# 8 More Bumps in the Road

In *Making Number Talks Matter* (Humphreys and Parker 2015, 163–175), we address several "bumps in the road" that teachers experience with Number Talks and suggest ways to get beyond those bumps. Here we discuss additional questions and concerns that teachers have raised over the past couple of years. Some deal with relatively minor matters that can nevertheless make a big difference, while others address thornier issues. We are delighted that teachers are pushing our thinking with these kinds of questions!

## Teacher Questions and Concerns

### *Should Number Talks align with the curriculum I'm teaching?*

No. Whatever the grade level, we always begin Number Talks with whole numbers, one operation at a time. We don't even consider moving to fractions or decimals within an operation until students can work strategically and confidently with whole numbers. And while this trajectory is the same for all grade levels, the travel time will vary widely, depending on the age, needs, and experience of your students.

We do this on purpose. We've all seen what happens when students are taught arithmetic as procedures to remember, and it's common knowledge that by middle school all too many of them struggle to keep the rules straight or apply them correctly. This only gets worse. Over time, the rote procedures betray students, leaving them without the safety net that understanding how numbers work can provide. Number Talks give us an opportunity to shore up these crumbling foundations so that mathematics becomes something to think about, not just something to remember.

A strong and flexible understanding of whole number operations and arithmetic properties is the foundation for understanding operations with rational numbers, negative numbers, expressions, and equations. Sun, Baldinger, and Humphreys (2018, 51), for example, point out the following:

> Teachers might worry that focusing on number sense is not closely tied to district or state standards that must be addressed in high school math classrooms. Flexible use of numbers, however, underlies key ideas in algebra, geometry, trigonometry, and calculus. For example, being able to decompose numbers in multiple ways will support students to more easily factor quadratics. A strong foundation in ratios is essential for making sense of trigonometric relationships in right triangles. Understanding multiplication as area can support understanding of Riemann sum. Number Talks also help students understand subtraction strategies as distance, which is critical in geometry.

That said, when Number Talks are a part of your classroom culture, there will be moments when an opportunity arises for an informal Number Talk that *is* related to your curriculum. In this video, for example, we see Cathy Young using area and perimeter as the context for a Number Talk in her fifth-grade classroom.

Cathy Young uses area and perimeter for a Number Talk with fifth graders.
http://sten.pub/dd48

Be on the lookout for times when a problem arises where it would take just a couple of minutes to stop and say, "How could we think about this?" and "Who thought about it differently?" Sometimes it's not even necessary to record. There are lots of places that whole number arithmetic occurs within the context of a lesson. You'll find several examples of this in *Making Number Talks Matter* (Humphreys and Parker 2015, 36).

The systemic damage to your students' reasoning about mathematics will take time to heal. So don't rush. And don't worry! As one teacher told us, Number Talks are all about "tending to the foundation."

## *Number Talks are going pretty well, but I worry because there are a few students I never hear from.*

It takes some students a long, long time to gain the confidence to share their thinking during Number Talks. Student engagement is such a complex issue that you need to keep tinkering with ideas and never give up on those students who are silent. Here are some things you can try:

- Hold small-group Number Talks. Give students a chance to try out their thinking in a small group, as some are more comfortable sharing their ideas this way. And for some, sharing in small-group Number Talks is the gateway to sharing in a large group. Heather Henry, a fifth-grade teacher in Washington State, holds an after-school club where she does *only* Number Talks. Many of her students who don't share in front of the whole class eagerly participate in the after-school program. Parents comment that the program has been a boost to their children's confidence and that they're more excited to go to school.

- Work behind the scenes to encourage participation. Talk with students one-on-one and let them know that you're interested in their thinking. Reassure them you won't call on them when they don't volunteer but explain that as their teacher, you want to know how they're thinking and you need to know how they're doing. Ask them to think about a problem, and then listen. It's good for them to be able to try their ideas out one-on-one. These conversations provide opportunities to interact with students around their ideas and offer encouragement. And this provides a chance to know what a student is learning and thinking.

  Often during these one-on-one check-ins, a student will share a surprising or interesting idea. When this happens, offer encouragement and invite the student to share his idea during the next Number Talk. If he indicates that he's not ready to share, ask if he's willing to let you share his idea. And if he is, be sure to ask if he wants to be identified. Most students are pleased by this invitation, and having their idea accepted or taken up by the class is a confidence booster.

- Try the quick formative assessment from *Making Number Talks Matter* (Humphreys and Parker 2015, 168) in order to know how every student is thinking about a problem.

- Try the revised Number Talk routine. The wait time ideas in Chapter 3 are designed to encourage greater participation.

While our goal is for every student to become comfortable sharing their thinking and engaging in small- and whole-group discussions, it's important to give individual students the time they need to be ready to share. This can be particularly challenging at the secondary level since the longer students have been turned off by mathematics, the harder we need to work to bring them back.

## *I've been doing Number Talks for a few months now, and sometimes I wonder if we're making progress.*

Time is precious. But without looking back, it's hard to know if we are making progress. Here are some suggestions for ways to analyze your practice with an eye towards progress (adapted from Friel 1992, 42):

- Videotape yourself. Notice what questions you ask, how you ask them, what you ask about, and how students respond. This can be illuminating and humbling; we both know this from experience!
- Keep a written log of your Number Talks with brief notes about what happened. Sometimes it's easier to see the progress we've made by looking back at how things were.
- Ask a colleague to watch one of your Number Talks. An eye in the back of the room can offer an important perspective that is impossible to have when you are alone. Ask your colleague to pay close attention to what questions you are asking and how the students respond.

## *What do you think about having students share strategies with a partner ahead of time so that everybody gets to talk?*

We haven't used this practice during Number Talks because we've seen it result in fewer answers and strategies being shared out with the whole class. It's *not* important that all students talk in any one Number Talk, but it *is* important that a few voices don't dominate. And it's important that, over time, every student who wants to will get a chance to share. We haven't had students turn and share their strategies, but if teachers really want all students to share, then they might (sparingly) have students put their heads together right after the Number Talk is over so that anyone who still wants to can share.

## *I'm confused about when to ask students to talk in small groups during Number Talks.*

You aren't alone! Jay Jahnsen, the teacher of the students we see in Chapter 1, "A High School Number Talk," told us after the lesson that he was surprised to see Cathy asking students to talk amongst themselves during the Number Talk:

Jay: "Is it okay to have kids talk?"
http://sten.pub/dd49

Jay didn't think having kids talk in small groups was part of the Number Talk routine. It's true that small-group discussions haven't been an explicit part of Number Talks, so it's no wonder that teachers might think they aren't okay.

In the past couple of years, though, we've become aware that with many Number Talks, students are willing to share their own ideas but show little interest in others'. So we've begun to focus on having students talk with each other to motivate more active student engagement with others' ideas, as Cathy was doing in the Number Talk in Chapter 1. This is new territory for us. One of the things we've written about is staying out of the way of kids' mathematical thinking by not leading or guiding their ideas. But this is different from using purposeful teaching moves to help students realize their new role as being part of a collaborative community. Students have well-established habits for how they interact in a math class, but listening to and building on each other's ideas are not typically among those habits. This means we need to be explicit about initiating students into new ways of interacting in math class.

So when *do* we ask them to talk with each other, and when *don't* we? We've gradually arrived at some principles that guide us in deciding when and when not to ask students to talk with each other. In this next video, Cathy talks about how she makes those decisions.

Cathy talks about how she decides when to have students talk to each other during Number Talks.
http://sten.pub/dd50

Having students talk with each other at these critical times during Number Talks is important. Talking can open up a space for students to put an idea into their own words in order to clarify (for themselves) what they understand and don't understand about someone else's idea. It can give them a chance to toss ideas around more privately before publicly making a conjecture or asking a question. It also gives students more opportunities to really listen to each other. Besides, everyone can benefit from the small-group discussions that ensue, even those who don't choose to talk.

But be careful. Cathy admits, in Chapter 1, that she overdid asking students to rephrase each other's thinking. She was in a unique situation, though, that won't be true for you: she had just this one chance to try to get a class talking to each other. You'll be able to work on stirring up interaction over time. And gradually, if you've worked at this and you're lucky, you'll find that the need to use these teaching moves will diminish as students increasingly take on a sense of shared ownership of the classroom discourse. Even though you'll need to use these moves less frequently over time, there will always be moments when you recognize something that you want kids to dig into. That's the time to have kids talk together.

## *Help! It's so hard not to say things like, "Great!" when a student has shared.*

This is hard for us, too! We've been trying to maintain a neutral stance for years, and yet we still catch ourselves saying things like, "Interesting!" or "Yes," even though we know that these simple comments can cause students to believe that we value some ideas more than others. Besides, if we respond with praise to every idea, the praise becomes meaningless.

Meaningless praise, though, is not necessarily benign. It teaches students to look to their teachers for validation. Indeed, a large body of research suggests that generic praise can keep students dependent on us (Boaler 2016; Kohn 1999). Peter H. Johnston (2012) opens his chapter "'Good Job!' Feedback, Praise, and Other Responses" with this quote from Norman Vincent Peale: "The trouble with most of us is that we would rather be ruined by praise than saved by criticism" (35). It's a sobering thought.

The trouble is, it's difficult to know what to do instead. We've tried saying, "Okay," "Thank you," and other similar responses, but even those can become habitual and meaningless. So, what to do? We wonder if pausing to give students a chance to respond (see Chapter 3, "Revising the Number Talk Routine: Rethinking Wait Time") might work better than saying anything at all. Whatever we decide, we tell students that even when their ideas "knock our socks off," we might not say, "Wow!" That's because we're trying hard to help them rely less on us to determine whether their ideas are sound and more on themselves. There's no getting around it: it's just darn hard to get ourselves out of the way, but Number Talks give us a chance to practice.

When students' ideas do "knock our socks off," we try to let them know, one-on-one, that we learned something from them or noticed something important about what they did. We offer encouragement but try to be specific about what was important and why.

## *I don't know what to do with the interesting ideas that come up when there is so little time during a Number Talk.*

A big part of the beauty of Number Talks is that students, on their own, uncover interesting mathematical ideas and ask intriguing questions. Yet deciding what to do with these ideas and questions in the moment can be tricky. There is no one right way to make these kinds of decisions. Sometimes the ideas or questions relate directly to the grade-level content, and then it's easy to go with the ideas when they arise. Sometimes, though, there is no way to extend the time, and exploring the ideas gets put off until another Number Talk. It can help to have a place to post students' questions or ideas that need further exploration. And sometimes, with our all-too-crowded grade-level expectations, even good ideas just get dropped. What's most fun for us, though, is when students won't let an idea or question go, and on their own seem to erupt into animated discussions or debates. When this happens, we see it as a positive sign that students are developing a sense of agency. Through their questions and ideas, students can become the generators as well as consumers of curriculum. And when something comes up that neither you nor your students want to let go, then it might just be time for an investigation, as we describe in *Making Number Talks Matter* (Humphreys and Parker 2015, 133–162).

## *Why do I do all the recording? Could I turn some of it over to students?*

We ask teachers to record during Number Talks for three main reasons: (1) clear communication, (2) accurate representation, and (3) precise mathematical notation. During recording, in a process that even teachers can find challenging, purposeful decisions need to be made about how to record each part of what a student says, in order to accurately represent and communicate that student's thinking. The recording must be mathematically accurate, but also should authentically communicate the student's thinking and be large enough to be seen across the room. This is too much to ask of students. It's hard enough to understand what someone has said, let alone record it accurately. Also, students are far more likely to impose their own reasoning processes on someone else's thinking, in essence recording their "take" on what the student has said. Students' recordings also frequently model incorrect mathematical notation. Number

Talks provide a brief time for students to think about how numbers work, and how they might make problems easy to solve mentally. So for all these reasons, we think it's better to have teachers do the recording.

## *I find myself getting ahead of my students and recording things they haven't yet said, even when I'm trying not to!*

It's good to catch ourselves getting ahead of students while recording their ideas, even if we notice too late. This is something we've struggled with, too. Recording only what students have already said communicates that we are genuinely curious about what they have to say. When we get ahead of them, though, we inadvertently send the message that we know what they've done and where they're going with an idea, which can rob them of experiencing the merit of their own ideas. We've also had many surprises when we *thought* we knew what a student was going to say. As with the first question, this is an issue of slowing things down weighed against the constant need to keep things moving.

## *What can I do when I find it hard to keep up while trying to record students' thinking?*

Slow down. While this advice might seem paradoxical, we've found that slowing down while we are recording gives students more time to think about what others are saying. If teachers have a hard time keeping up, it's a fair bet that students do as well. Slowing down helps others anticipate what someone is doing and gives space to consider the ideas being put forward.

## *What do you think about using journals or individual whiteboards during Number Talks so I can see all kids' work?*

One of the biggest problems teachers face is how entrenched standard algorithms are in students' thinking about computation. So while journals or individual whiteboards are worthwhile in other settings, their use during Number Talks can make it nearly impossible to liberate students from their dependence on rote procedures. When a pencil and paper are nearby, students often start writing even before they begin thinking. The focus during Number Talks must always be on finding efficient, personal ways to solve computation problems *mentally*.

It's understandable, though, to feel the need to know how everyone is thinking, not just the students who have spoken that day. For this reason, we recommend doing a quick formative assessment once a month or so to help you get a handle on where each of your students is (Humphreys and Parker 2015, 168).

## *I don't know how to find time for Number Talks every day, so I'm only doing them on Monday, Wednesday, and Friday.*

Congratulations for finding time to do Number Talks three days a week! If you have any flexibility with time, you might find that doing them on three consecutive days is even more productive. Students build on each other's ideas during Number Talks, and it can be hard—for us and our students—to remember what happened two days ago. If you're unable to shift days, don't worry. Number Talks three times a week will make a big difference in students' mathematical understanding and dispositions.

## *It's hard to end a Number Talk after fifteen minutes when there are still students who want to share.*

Yes, sometimes Number Talks could take over the entire math lesson if you let them. But we typically try to keep them to fifteen minutes so that we can do them almost every day. Some teachers like to use a timer to remind them when time is up. And even though you might have students who still want to share at the end of the fifteen minutes, you can acknowledge that you didn't get to everyone and express the hope that those who didn't get to share will do so next time. It can also help to informally keep track of who was still waiting to share when the Number Talk is over.

## *With so much else to do and student participation lagging, I've started doing Number Talks less and less.*

If few students are participating in a Number Talk, it can feel like the class time isn't very well spent, so it's natural to find yourself putting them off. We all encounter ups and downs with student participation, and when participation ebbs, it can be hard to see how students are benefiting. Number Talks are a "long game," though, so the fulfillment of their promise depends on keeping them going. If we don't give up—if we keep them going—then more students gradually will participate, the real value of Number Talks will become more apparent, and you will feel encouraged and have more confidence to keep them going.

Lots of things impact student participation during Number Talks. Some students are shy. Many are convinced they're no good at math—and sometimes you might have a classful of students who generally lack confidence in themselves as mathematical thinkers. Some students like to think more deeply about ideas before speaking; others want time to compose what to say before they're willing to talk (these show why wait time is so important). Some fear being wrong or letting others know they don't understand.

In the video that follows, Tara talks about the impact that fear of being wrong has had on her geometry students.

Tara comments that her students are beaten down by math.
http://sten.pub/dd51

The belief that being good at math means being fast at getting right answers has taken a real toll on our students' willingness to engage and take risks, as Ilana Horn (2017) points out:

> The cultural power of mathematical smartness makes math class a minefield of social risk. For example, students often fear that any visible signs of struggle not only implicate their precarious understanding of the topic at hand but also compromise their general sense of competence . . . This preoccupation with mistakes alone can impede students from participating in math class. (60)

Many things inhibit students' willingness to participate, so when you get discouraged, it might help to remember that it takes time for students to rebuild confidence in themselves. Here are some things to try when you find yourself in one of those valleys:

- Success is a great motivator—go back to Dot Talks for a while.
- Make participation a priority for a while and actively stir things up, as Cathy did in Chapter 1.
- Get serious about wait time (see Chapter 3).
- Try some of the nudging ideas in Chapter 5.

- And, for your own encouragement, enlist the aid of a colleague. Having even one colleague to work with can make it easier to dedicate the space and time for Number Talks. In this video, Tara talks about how much she needed Melissa as a colleague when they were first starting to implement Number Talks in their classrooms.

Tara talks about why it's important to have a colleague with whom to work.
http://sten.pub/dd52

So find a colleague, share your concerns, and work together to keep Number Talks going. There's too much that is lost in letting them fall by the wayside.

## *Our question for you: How are you and your students feeling about Number Talks?*

At the 2017 NCTM Research Conference, a session called "Joy: The Zeroeth Mathematical Practice" (Parks and Wager 2017) caught our attention. Weaving together ideas from physics, mathematics education research, and *The Book of Joy* (Dalai Lama and Desmond Tutu 2016), Amy Parks and Anita Wager characterized joy, which is missing from so many mathematics classrooms, as a fundamental principle that should be the foundation for all math teaching and learning.

Parks and Wager (2017) helped remind us why more joy is needed in mathematics teaching and learning. Some of their ideas, in particular, struck a chord with us: that joy and sadness are not opposites, but joy and disengagement are; that joy is an equity issue; and that when true mathematical thinking is happening, joy is involved. We believe that Number Talks, with their emphasis on inviting all voices and actively engaging all students in mathematical sense making, can awaken joy in mathematics classrooms. So, as we're all in the midst of working with the ideas in this book to make Number Talks more vibrant and meaningful, let's remember to cultivate joy.

We're aware that we've only scratched the surface when it comes to teachers' questions about Number Talks. We've been thinking about these ideas for years, and we still have many questions of our own. We've come to think of Number Talks as a journey, one in which teachers will continue to bump into new obstacles and discover new horizons. Over time, though, the obstacles diminish and new horizons expand.

# Afterword

*The instructor should never forget the mutual relationship between horse and rider. The rider's faults in seat and in guidance are reflected in the movements of the horse, while shortcomings of the horse's training make it difficult for the rider to sit and perform exercises correctly.*

—ALOIS PODHAJSKY, *THE COMPLETE TRAINING OF HORSE AND RIDER IN THE PRINCIPLES OF CLASSICAL HORSEMANSHIP*

## Thoughts from Cathy

You're probably wondering what this epigraph by the director of the famed Spanish Riding School in Vienna, Austria, has to do with Number Talks. But this is something that has been sitting at the back of my mind for many years. When I was just beginning my teaching career, I took lessons in dressage for a short time. One day my instructor, who knew I was a teacher, used a classroom metaphor that he thought would help me become a better rider. His analogy stayed with me, and I think it has helped me understand why it can be so hard to break the stranglehold that our mathematics cultural tradition has on teaching.

My riding instructor characterized good riding as a *collaboration* between horse and rider—between two living things—just as good teaching is a collaboration between a teacher and her students. A rider communicates with her horse through "aids" (hands, legs, seat), and together, horse and rider negotiate a shared understanding of what each aid means.

Collaboration is a reciprocal process. Rouen, a sympathetic and patient Arabian gelding, taught me to ride. When I applied an aid correctly, Rouen let me know, his quick responses telling me that I had done something right. Then, when I repeated the aid, Rouen would again respond. It took time, but I was improving. Rouen, the expert, and I, the novice, were working together.

My instructor also talked to me about a young, "green" horse he was training. He described how, when the young horse responded correctly to an aid, he released the aid immediately. Just like me, the novice horse began to learn what to pay attention to.

But what if *I* was on the green horse? We would both, as novices, be learning together. We would have to work out how to communicate with each other and certainly would give each other confusing and mixed messages along the way. It would take a lot more patience, guidance, and time, and I can easily see myself wanting to give up on the whole thing (and can only imagine how the poor horse would feel).

So what's the connection with Number Talks? During Number Talks, both students and teachers are novices in these new ways of interacting. But when teachers, imagining a classroom filled with sense-making students and lively discourse, ask students about their ideas (and try to hold back from explaining), students are often not yet able to respond in hoped-for ways. Everybody feels off balance, and teachers don't get the encouraging responses they need to keep them going.

But what are we asking of students? Here are ways that Number Talks shift the classroom landscape for students (Table 1).

**TABLE 1**
Shifts for Students During Number Talks

| FROM | TOWARDS |
|---|---|
| Remembering | Reasoning |
| Focusing on the answer | Focusing on how they get their answers |
| Doing a problem "the teacher's way" | Doing a problem in a way that makes sense to them |
| Expecting the teacher to tell them whether their answers are right or wrong | Expecting their teacher to ask them to verify the answers for themselves |
| Trying to avoid mistakes | Seeing mistakes as opportunities to learn |
| Seeing confusion as bad | Getting comfortable with cognitive dissonance |
| Reproducing their teacher's method and asking the teacher when they get stuck | Listening to their classmates' methods and asking *them* questions |
| Thinking there is one way to solve a problem | Knowing there are many ways to solve problems |
| Relying on the teacher to direct discourse | Jumping in to share responsibility for discourse |

*Source:* Adapted from Humphreys (2016).

These are seismic shifts in what students think math *is*, what they think teachers (and their classmates) are supposed to do, and what it means to be smart. And the hardest part for teachers is to respond immediately, as Rouen did, whenever students make incremental shifts, and to do so under the considerable strain of the new things they themselves are learning (Table 2).

**TABLE 2**
Shifts for Teachers During Number Talks

| FROM | TOWARDS |
|---|---|
| Asking questions you already know the answers to | Asking questions you're genuinely curious about |
| Focusing on answers | Focusing on reasoning |
| Being the "primary explainer" | Letting kids do the thinking |
| Listening *for* a particular response | Listening *to* students |
| Demonstrating how to solve problems | Finding out how students solve problems |
| Verifying correct answers | Expecting students to verify their own answers |
| Knowing ahead of time where a lesson will go | Responding flexibly to where students' thinking leads |
| Being in charge of classroom discourse | Sharing responsibility with students for classroom discourse |

*Source:* Adapted from Humphreys (2016).

It's common in mathematics education circles to emphasize the importance of establishing a class culture where students feel safe to share and discuss their ideas. *Establishing*, though, which implies a one-way, teacher-directed street, is a myth. How can teachers "establish" safety? Or "establish" students' beliefs about mistakes, or what it means to be smart in math? Clearly, we can't. This is where Podhajsky's comment applies. He reminds us that as we try to change our own practice, we're in a mutual, responsive relationship with our students. All we can do is keep at it, sharing our discomfort, making our reasons as clear as possible, and forging ahead in a truly collaborative effort.

# Appendix A

# Different Ways to Ask Questions During Number Talks

## Inviting Answers

- It looks like we're ready to hear solutions.
- Is anybody willing to raise their hand and share what they think the answer might be?
- Who's willing to share their answer?
- Is anyone willing to say what they think the answer might be?
- Are there any other answers?

## Inviting Strategies

- Is anyone willing to share how they figured it out?
- Is anyone willing to convince us that your answer makes sense by telling us how you thought about it?
- Is anybody who thought about it differently willing to share what they did?
- Who did it a different way?
- Other ways of thinking about this?

## Pressing for Why

- Why did you choose to start with . . . ?
- Can you tell us why you . . . ?
- I notice you . . . Could you tell us more about this?
- Can you tell us how that made it easier for you?
- How did you know you needed to . . . ?

# Appendix B

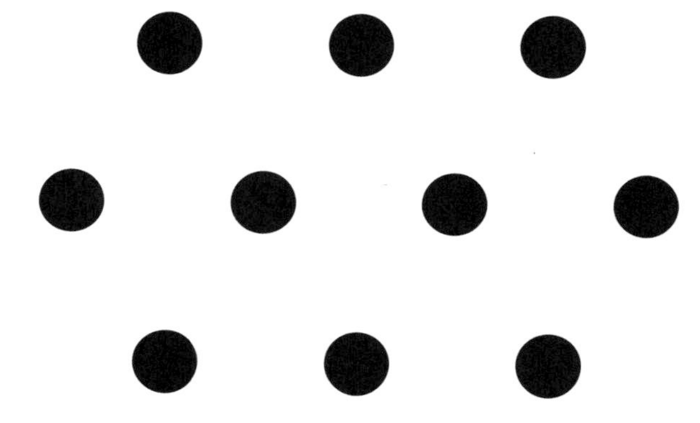

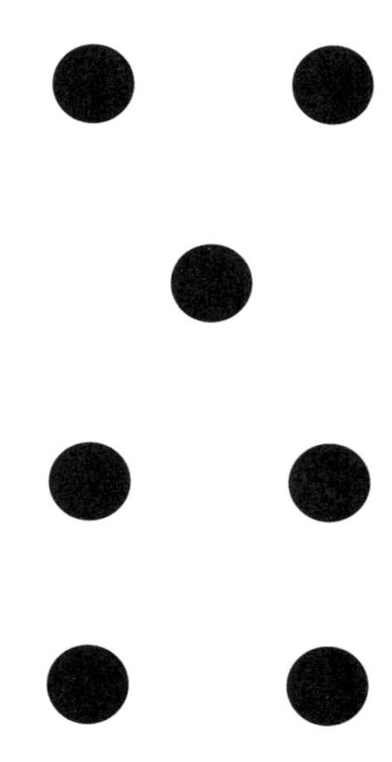

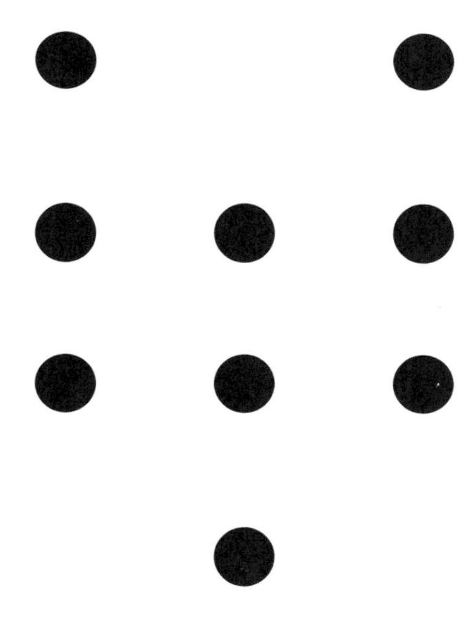

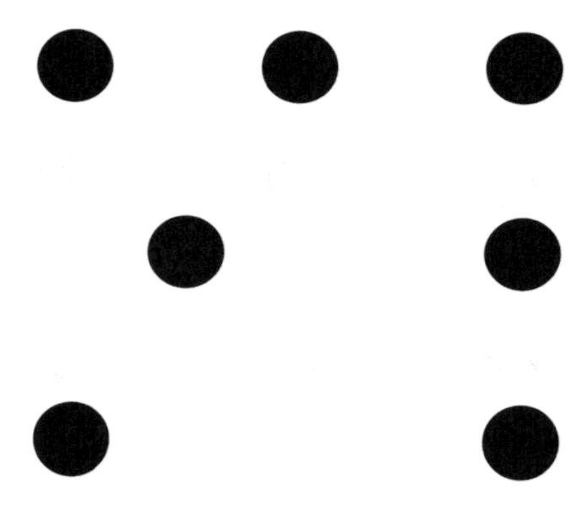

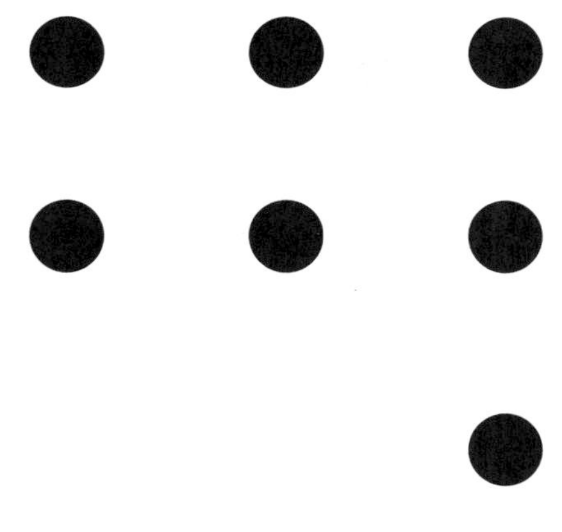

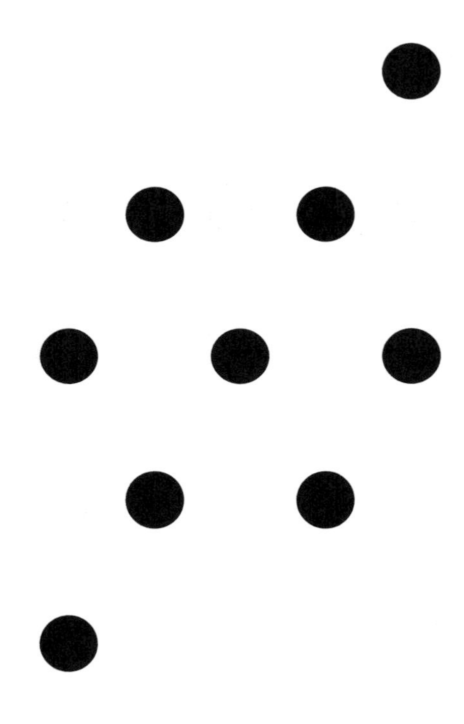

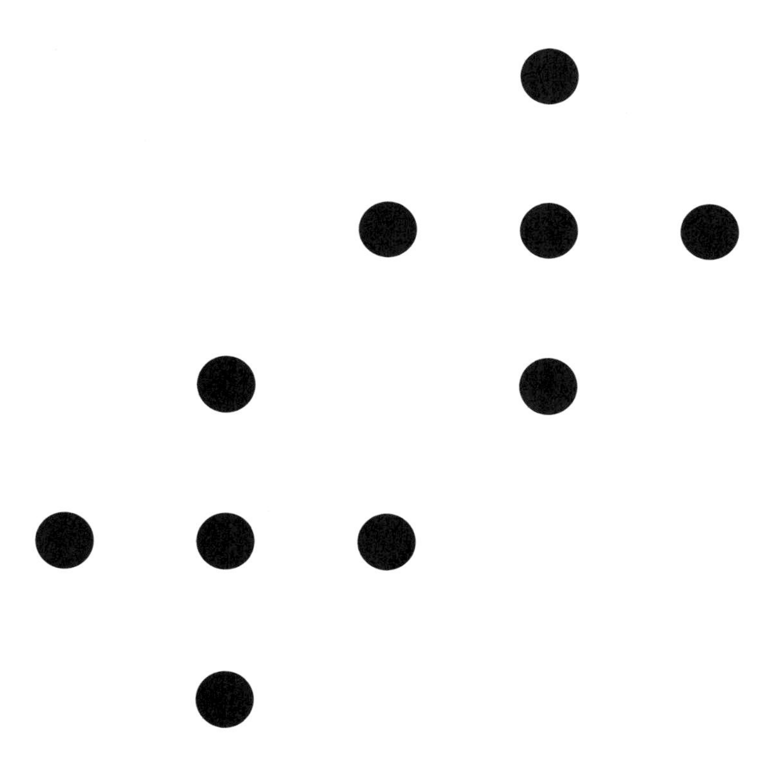

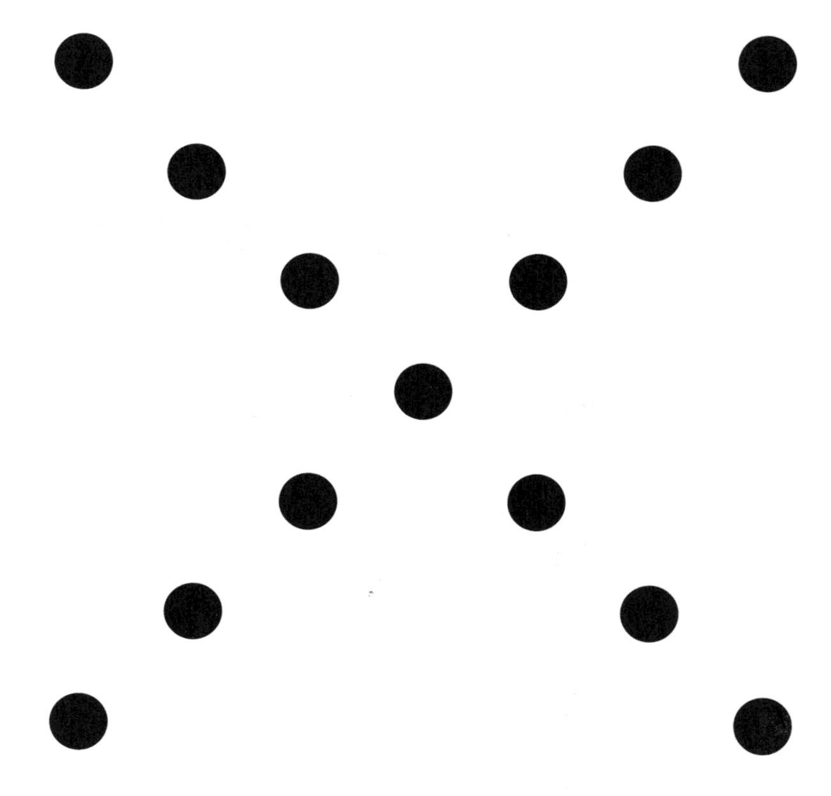

# Appendix C

## Revised Number Talk Routine in a Nutshell

1. Students put paper and pencils away and put their closed hands on their chests.
2. Write the problem (horizontally on the board).
3. WAIT while students think about the problem.
4. Early on, remind students not to agree or disagree with answers that are offered.
5. Ask students to put their thumb up if they're willing to share an answer.
6. Record only answers on the board and continue to ask for different answers.
7. Ask students to put their thumbs up again if they're willing to share what they did; then WAIT.
8. Ask which answer they're defending, record the student's thinking, and then WAIT.
9. Ask if anyone solved it differently; then WAIT.
10. Ask students to look across the strategies for wonderings or noticings; then WAIT.
11. Thank students and end the Number Talk.

# Video Contents

Subtraction of mixed numbers: A high school Number Talk.
http://sten.pub/dd01
P. 8

Students restate Troy's strategy.
http://sten.pub/dd02
P. 10

Paul uses the open number line to subtract.
http://sten.pub/dd03
P. 10

Frosty.
http://sten.pub/dd04
P. 11

It's important for people to speak up when we don't understand, like Frosty has just done.
http://sten.pub/dd05
P. 12

Frosty restates his question.
http://sten.pub/dd06
P. 12

"Why are you guys getting so many different answers?"
http://sten.pub/dd07
P. 12

Being wrong is part of learning math.
http://sten.pub/dd08
P. 13

"Bad at math."
http://sten.pub/dd09
P. 14

Frosty helps decide how to process the discussion.
http://sten.pub/dd10
P. 14

The Number Talk ends in an unexpected way.
http://sten.pub/dd11
P. 15

Frosty describes his strategy.
http://sten.pub/dd12
P. 17

At the end of her third-grade Number Talk, Hailey invites a defense of the other answers.
http://sten.pub/dd13
P. 20

Tiancum offers 45 but defends 55.
http://sten.pub/dd14
P. 21

Megan figures out what happened.
http://sten.pub/dd15
P. 22

Seth corrects his own mistakes.
http://sten.pub/dd16
P. 22

Danni corrects her own mistake.
http://sten.pub/dd17
P. 23

Mica misapplies a subtraction strategy by adding the numbers.
http://sten.pub/dd18
P. 28

Cathy uses a multiplication Number Talk to examine the distributive property.
http://sten.pub/dd19
P. 30

Jamie asks students what they notice or wonder.
http://sten.pub/dd20
P. 39

Jay presses Paul to explain how he knew that 74 – 25 is 49.
http://sten.pub/dd21
P. 47

Hailey "presses" on an important part of Danni's strategy.
http://sten.pub/dd22
P. 48

Hailey focuses on why Isa added 4 to both numbers.
http://sten.pub/dd23
P. 48

Nisha presses Logan for
elaboration and justification.
http://sten.pub/dd24
P. 49

Listening to students.
http://sten.pub/dd25
P. 51

Ruth is confused about Aurora's explanation.
http://sten.pub/dd26
P. 55

Logan forgets what he did.
http://sten.pub/dd27
P. 55

Tara talks about learning to ask
open-ended questions.
http://sten.pub/dd28
P. 56

Tara talks about how hard it is for students to
get beyond standard algorithms.
http://sten.pub/dd29
P. 58

Hailey gathers answers for 62 – 29.
http://sten.pub/dd30
P. 58

Colson uses the standard algorithm for 63 – 30.
http://sten.pub/dd31
P. 59

Emily and Hailey wonder about the number of answers and the "add one" strategy.
http://sten.pub/dd32
P. 59

Emily, literally, "adds instead."
http://sten.pub/dd33
P. 59

Hailey, Ruth, Cathy, and Emily discuss what happens and what to do when students are following procedures they've been taught.
http://sten.pub/dd34
P. 59

Grace uses a commonly taught alternative method for multiplying 32 x 12.
http://sten.pub/dd35
P. 60

Hailey also uses the distributive property.
http://sten.pub/dd36
P. 60

Deja makes it easy to solve 32 x 12.
http://sten.pub/dd37
P. 60

Starting the Dot Talk.
http://sten.pub/dd38
P. 65

Emily is subitizing.
http://sten.pub/dd39
P. 66

Ruth records Addie's way of seeing.
http://sten.pub/dd40
P. 67

Elijah says "diamond" and Ruth records, saying, "rhombus."
http://sten.pub/dd41
P. 68

Micah struggles to describe two dots and Ruth casually introduces the term *vertical*.
http://sten.pub/dd42
P. 68

Ruth records Addie and Mica's ways of seeing.
http://sten.pub/dd43
P. 69

Ruth, Hailey, Emily, and Cathy discuss how we record during Dot Talks, and why.
http://sten.pub/dd44
P. 69

Damien explains what he sees as Ruth listens without looking at him.
http://sten.pub/dd45
P. 71

Ruth looks away to nudge students to use words for what they see.
http://sten.pub/dd46
P. 71

Gloria explains what she sees.
http://sten.pub/dd47
P. 72

Cathy Young uses area and perimeter for a Number Talk with fifth graders.
http://sten.pub/dd48
P. 82

Jay: "Is it okay to have kids talk?"
http://sten.pub/dd49
P. 85

Cathy talks about how she decides when to have students talk to each other during Number Talks.
http://sten.pub/dd50
P. 85

Tara comments that her students are beaten down by math.
http://sten.pub/dd51
P. 90

Tara talks about why it's important to have a colleague with whom to work.
http://sten.pub/dd52
P. 91

# References


Boaler, Jo. 2016. *Mathematical Mindsets: Unleashing Students' Potential Through Creative Math, Inspiring Messages and Innovative Teaching.* San Francisco, CA: Jossey-Bass.

California State Department of Education. 1988. *Mathematics Model Curriculum Guide: Kindergarten Through Grade Eight.* Sacramento, CA: California State Department of Education.

Carpenter, Thomas P., Megan L. Franke, Nicholas C. Johnson, Angela Chan Turrou, and Anita A. Wager. 2017. *Young Children's Mathematics: Cognitively Guided Instruction in Early Childhood Education.* Portsmouth, NH: Heinemann.

Dalai Lama and Desmond Tutu. 2016. *The Book of Joy.* New York: Avery.

Darling-Hammond, Linda, and Frank Adamson. 2010. *Beyond Basic Skills: The Role of Performance Assessment in Achieving 21st Century Standards of Learning.* Stanford, CA: Stanford Center for Opportunity Policy in Education (SCOPE), Stanford University School of Education.

Friel, Susan N. 1992. "The Role of Reflection in Teaching." *Arithmetic Teacher* 1:40–42.

Hiebert, James. 1999. "Relationships Between Research and the NCTM Standards." *Journal for Research in Mathematics Education* 30 (1): 3–19.

Hiebert, James, Thomas Carpenter, Elizabeth Fennema, Karen Fuson, Diana Wearne, Hanlie Murray, Alwyn Olivier, and Piet Human. 1997. *Making Sense: Teaching and Learning Mathematics with Understanding.* Portsmouth, NH: Heinemann.

Horn, Ilana Seidel. 2017. *Motivated: Designing Math Classrooms Where Students Want to Join In.* Portsmouth, NH: Heinemann.

Humphreys, Cathy. 2016. "Number Talks in High School." *New England Mathematics Journal* 48 (1): 28–39.

Humphreys, Cathy, and Ruth Parker. 2015. *Making Number Talks Matter: Developing Mathematical Practices and Deepening Understanding, Grades 4–10.* Portland, ME: Stenhouse.

Johnston, Peter H. 2004. *Choice Words: How Our Language Affects Children's Learning.* Portland, ME: Stenhouse.

———. 2012. *Opening Minds: Using Language to Change Lives.* Portland, ME: Stenhouse.

Kazemi, Elham, Megan Franke, and Magdalene Lampert. 2009. "Developing Pedagogies in Teacher Education to Support Novice Teachers' Ability to Enact Ambitious Instruction." In *Crossing Divides: Proceedings of the 32nd Annual Conference of the Mathematics Research Group of Australasia*. Vol. 1, ed. B. B. R. Hunter and T. Burgess. Palmerston North, New Zealand: Massey University College of Education.

Kohn, Alfie. 1999. *Punished by Rewards: The Trouble with Gold Stars, Incentive Plans, A's, Praise, and Other Rewards*. Boston: Houghton Mifflin.

Lampert, Magdalene. 2001. *Teaching Problems and the Problems of Teaching*. New Haven, CT: Yale University Press.

NGA/CCSSO (National Governors Association Center for Best Practices & Council of Chief State School Officers). 2010. *Common Core State Standards for Mathematics*. Washington, DC: National Governors Association Center for Best Practices & Council of Chief State School Officers.

Nystrand, Martin, Lawrence Wu, Adam Gamoran, Susie Zeiser, and Daniel Long. 2003. "Questions in Time: Investigating the Structure and Dynamics of Unfolding Classroom Discourse." *Discourse Processes* 35 (2): 135–198.

Parks, Amy Noelle, and Anita A. Wager. 2017. "Joy: The Zeroeth Mathematical Practice." NCTM Research Conference. San Antonio, TX, April 3–5.

Rowe, Mary Budd. 1986. "Wait Time: Slowing Down May Be a Way of Speeding Up." *Journal of Teacher Education* 37:43–50.

Podhajsky, Alois. 1967. *The Complete Training of Horse and Rider in the Principles of Classical Horsemanship*. North Hollywood, CA: Wilshire Books.

Schoenfeld, Alan. 2017. "What Have We Learned over the Past 60 Years, and Where Might We Be Going? A Biased History and a View of What Counts in Mathematics Teaching and Learning." Presentation at the California Mathematics Council, Northern Section Conference at Asilomar, Pacific Grove, CA, December 1–3.

Sun, Kathy L., Erin E. Baldinger, and Cathy Humphreys. 2018. "Number Talks: Gateway to Sense Making." *Mathematics Teacher* 112 (September): 48–54.

Thompson, Alba G., Randolph A. Philipp, Patrick W. Thompson, and Barbara A. Boyd. 1994. "Calculational and Conceptual Orientations in Teaching Mathematics." In *Professional Development for Teachers of Mathematics*, ed. Douglas B. Aichele and Arthur F. Coxford, 79–92. Reston, VA: National Council of Teachers of Mathematics.

Wood, Terry. 1998. "Alternative Patterns of Communication in Mathematics Classes: Funneling or Focusing." In *Language and Communication in the Mathematics Classroom*, ed. Heinz Steinbring, Maria G. Bartolini Bussi, and Anna Sierpinska, 167–178. Reston, VA: National Council of Teachers of Mathematics.

Youcubed at Mathematical Perspectives. Kathy Richardson https://www.youtube.com/watch?v=Ihz-0pGmhLI&t+13s.

Youcubed at Stanford University. "Cathy Humphreys Teaching a Number Talk." https://www.youcubed.org/resources/cathy-humphreys-teaching-number-talk/.

Youcubed at Stanford University. "Mistakes." https://www.youcubed.org/?s=mistakes.

# Index